Essential Maths for Students

Foundation Maths

Anthony Croft
and
Robert Davison

D0453048

Longman
Scientific &
Technical

Longman Scientific & Technical
Longman Group Limited
Longman House, Burnt Mill, Harlow
Essex CM20 2JE, England
and Associated Companies throughout the world

©Longman Group Limited 1995

All rights reserved; no part of this publication may be
reproduced, stored in any retrieval system, or transmitted in
any form or by any means, electronic, mechanical,
photocopying, recording, or otherwise without either the prior
written permission of the Publishers or a licence permitting
restricted copying in the United Kingdom issued by the
Copyright Licensing Agency Ltd., 90 Tottenham Court Road,
London W1P 9HE.

First published 1995

British Library Cataloguing in Publication Data
A catalogue entry for this title is available from the British Library.

ISBN 0–582–23185–X

Set by 16 in 10/12pt Times
Produced by Longman Singapore Publishers (Pte) Ltd.
Printed in Singapore

Contents

Preface

Today, a huge variety of disciplines require their students to have knowledge of certain mathematical tools in order to appreciate quantitative aspects of their subjects. At the same time, higher education institutions have widened access so that there is much greater variety in the mathematical profile of the student body. Many of these students are returning to education after many years in the workplace or at home bringing up families. One consequence has been that lecturers can no longer assume that a firm bedrock of mathematical knowledge and competence has been established. In addition, students are finding themselves inadequately prepared for the mathematical demands placed upon them. In some institutions, a remedy has been access or foundation course provision which prepares students to make the transition to a higher education course. In others, numeracy and mathematics learning centres have been established to provide support and encouragement.

Foundation Maths has been written to pave the way into higher education for those students who have not specialized in mathematics at *A* level. It is intended for non-specialists who need some but not a great deal of mathematics as they embark upon their courses of higher education. It takes students from around the lower levels of GCSE to a standard which will enable them to participate fully in a degree or diploma course. It is ideally suited for foundation and access courses in mathematics and for those who wish to enter a wide range of courses such as marketing, business studies, management, science, engineering, social science, geography, combined studies and design. It will also be useful for those who lack confidence and need careful, steady guidance in mathematical methods. Even for those whose mathematical expertise is already established, the book will be a helpful revision and reference guide. The style of the book also makes it suitable for those who wish to engage in self-study or distance learning.

We have tried throughout to adopt an informal, user-friendly approach and have described mathematical processes in everyday language. Mathematical ideas are usually developed by example rather than by formal proof. This reflects our experience that students learn better from examples than from abstract development. Where appropriate, the examples contain a great deal of detail so that the student is not left wondering how one stage of a calculation leads to the next. In

Foundation Maths, objectives are clearly stated at the beginning of each chapter, and key points and formulae are highlighted throughout the book. Self-assessment questions are provided at the end of most sections. These test understanding of important features in the section and answers can be checked by referring back. These are followed by exercises; it is essential that these are attempted as the only way to develop competence and understanding is through practice. Detailed solutions to these exercises are given at the back of the book and should be consulted only after the exercises have been attempted. A further set of test and assignment exercises is given at the end of each chapter. These are provided so that the tutor can set regular assignments or tests throughout the course. Solutions to these are not provided.

In order to keep the size of the book reasonable, we have endeavoured to include topics which we think are most important, cause the most problems for students and have the widest applicability. Inevitably compromise has been necessary. However, recent experience has shown us that the root of many of the problems occurring in mathematical education today is the lack of knowledge of algebraic manipulation. So many staff exclaim that if only their students could master algebra, all of the other techniques would fall into place. A large proportion of students now entering higher education have had little opportunity to develop algebraic skills to the standard needed in applications. With this in mind we have placed great emphasis on algebra in the book, with new work being gradually introduced over several chapters.

The best strategy for those using the book would be to read through each section, carefully studying all of the worked examples and solutions. Many of these solutions develop important results needed later in the book. It is then a good idea to cover up the solution and try to work the example again for yourself. It is only by actually doing the calculation for yourself that the necessary techniques will be mastered. At the end of each section attempt the self-assessment questions. If you cannot answer these, read the previous few pages again to find the answers in the text. Then attempt the exercises, regularly checking your solutions with those given at the back of the book.

In conclusion, remember that learning mathematics takes time and effort. Carrying out a large number of exercises allows you to experience a greater variety of problems, thus building your expertise and confidence. Armed with these you will be able to tackle more unfamiliar and demanding problems that arise in other aspects of your course. We hope you find *Foundation Maths* useful and wish you the very best of luck.

A. Croft and R. Davison

Mathematical symbols

+   plus
−   minus
±   plus or minus
×   multiply by
·   multiply by
÷   divide by
=   is equal to
≡   is identically equal to
≈   is approximately equal to
≠   is not equal to
>   is greater than
≥   is greater than or equal to
<   is less than
≤   is less than or equal to
∈   is a member of set
ℕ   set of natural numbers
ℤ   set of integers
ℝ   set of real numbers
ℚ   set of rational numbers
∴   therefore
∞   infinity
e   the base of natural logarithms (2.718...)
ln natural logarithm
log logarithm to base 10
∑   sum of terms
∫   integral
$\frac{dy}{dx}$ derivative of y with respect to x

1 Arithmetic

Objectives

This chapter

- classifies numbers

- explains the BODMAS rule for adding, subtracting, multiplying and dividing arithmetical expressions

- shows how to write a number to a given number of significant figures

- shows how to write a number to a given number of decimal places

- explains the terms 'highest common factor' and 'lowest common multiple'

1.1 The number line

Numbers are classified into various groups or **sets** as they are often called. The **natural numbers** are 1, 2, 3, 4, 5, ... and so on. This set is denoted by $\mathbb{N}$. The numbers belonging to a set are usually enclosed in curly brackets. Hence we write

$$\mathbb{N} = \{1, 2, 3, 4, ...\}$$

We use the symbol $\in$ to denote 'belongs to'. Hence, since 2 belongs to the set $\mathbb{N}$ we write $2 \in \mathbb{N}$.

We use ... to mean that the numbers continue.

The **integers** are given by the set $\mathbb{Z}$:

$$\mathbb{Z} = \{..., -4, -3, -2, -1, 0, 1, 2, 3, 4, ...\}$$

The natural numbers are also known as the **positive integers**. An alternative notation to $\mathbb{N}$ is $\mathbb{Z}^+$, that is

$$\mathbb{Z}^+ = \{1, 2, 3, 4, 5, ...\}$$

The **negative integers**, denoted by $\mathbb{Z}^-$, are $-1, -2, -3, -4, -5, ...$,

$$\mathbb{Z}^- = \{-1, -2, -3, -4, ...\}$$

Note that the integers, $\mathbb{Z}$, comprise the natural numbers, the negative integers, and zero.

A number can belong to more than one set. For example, -3 belongs to $\mathbb{Z}^-$ and $\mathbb{Z}$, 3 belongs to $\mathbb{N}$, $\mathbb{Z}^+$ and $\mathbb{Z}$.

Rational numbers have the form p/q where p and q are integers, with $q \neq 0$. The value of q is not allowed to be zero because division by zero is not defined. Examples of rational numbers include $\frac{3}{7}$, $\frac{101}{93}$, $-\frac{7}{9}$ and $\frac{2}{1}$. The set of rational numbers is denoted by $\mathbb{Q}$. Note that all the integers are included in the set of rational numbers.

Finally, there are **irrational numbers**. Irrational numbers are those numbers which cannot be expressed in the form p/q. Examples include π and $\sqrt{2}$.

Any number can be represented by a point on a number line as given in Figure 1.1. The set of all numbers is represented by $\mathbb{R}$. The set $\mathbb{R}$ includes all the rational numbers and all the irrational numbers. Collectively they are known as the set of real numbers.

The symbol $\neq$ means 'is not equal to'

Figure 1.1
The number line

Greater than and less than

We use the symbol $>$ to mean 'greater than'. For example, $6 > 5$ and $-4 > -5$ are both true statements. If $x > y$ then x is to the right of y on the number line. The symbol $\geq$ means 'greater than or equal to', so, for example, $10 \geq 8$ and $8 \geq 8$ are both true.

The symbol $<$ means 'less than'. We may write $4 < 5$ for example. If $x < y$ then x is to the left of y on the number line.

Finally $\leq$ means 'less than or equal to'. Hence $6 \leq 9$ and $6 \leq 6$ are both correct statements.

Intervals

We often need to represent intervals on the number line. There are various ways of doing this.

(a) *The closed interval*   An interval which includes its end-points is called a closed interval. All the numbers from 1 to 3, including both 1 and 3, comprise a closed interval and this is denoted using square brackets, [1,3]. Any number in this closed interval must be greater than or equal to 1, and also less than or equal to 3. Thus if x is any number in the interval, then $x \geq 1$ and also $x \leq 3$. We write this compactly as $1 \leq x \leq 3$. Finally we need to show explicitly that x is any number on the number line, and not just a rational number or an integer for example. So we write $x \in \mathbb{R}$. Hence the interval [1,3] can be expressed as:

$$\{x : x \in \mathbb{R}, 1 \leq x \leq 3\}$$

This means the set contains all the numbers x with x a real number and $x \geq 1$ and $x \leq 3$.

(b) *The open interval*   Any interval which does not include its end-points is called an open interval. For example, all the numbers from 1 to 3, but excluding 1 and 3, comprise an open interval. Such an interval is denoted using round brackets, (1,3). The interval may be written using set notation as

$$\{x : x \in \mathbb{R}, 1 < x < 3\}$$

We say that x is **strictly greater** than 1, and **strictly less** than 3, so that the values of 1 and 3 are excluded from the interval.

(c) *The semi-open or semi-closed interval*   An interval may be open at one end and closed at the other. Such an interval is called semi-open or, as some authors say, semi-closed. The interval (1,3] is a semi-open interval. The square bracket next to the 3 shows that 3 is included in the interval; the round bracket next to the 1 shows that 1 is not included in the interval. Using set notation we would write

$$\{x : x \in \mathbb{R}, 1 < x \leq 3\}$$

When marking intervals on the number line there is a notation to show whether or not the end-point is included. We use • to show that an end-point is included (i.e. closed) while ○ is used to denote an end-point that is not included (i.e. open).

Worked examples

1.1   Describe the interval $[-3, 4]$ using set notation and illustrate it on the number line.

Solution   The interval $[-3, 4]$ is given in set notation by

$$\{x : x \in \mathbb{R}, -3 \leq x \leq 4\}$$

Figure 1.2 illustrates the interval.

Figure 1.2
The interval [−3,4]

1.2   Describe the interval (1,4) using set notation and illustrate it on the number line.

Solution   The interval (1,4) is expressed as

$$\{x : x \in \mathbb{R}, 1 < x < 4\}$$

Figure 1.3 illustrates the interval on the number line.

Figure 1.3
The interval (1,4)

Self-assessment questions 1.1

1. No number can be both a natural number and an irrational number. True or false?
2. No number can be both a rational number and an irrational number. True or false?
3. Explain the meaning of (a) a natural number, (b) an integer, (c) a rational number, (d) an irrational number and (e) a real number.
4. Explain why all natural numbers are also rational numbers. Is the contrary true?

Exercise 1.1

1. Describe the following intervals using set notation. Draw the intervals on the number line

   (a) [2,6]   (b) (6,8]   (c) (−2, 0)
   (d) [−3,−1.5)

2. State whether the following numbers are members of (i) $\mathbb{Z}$, (ii) $\mathbb{Z}^+$, (iii) $\mathbb{Z}^-$, (iv) $\mathbb{N}$.

   (a) 3 (b) −3 (c) 3.2 (d) 11 (e) 0.01

3. Which of the following are true?

   (a) 8 > 3 (b) −3 < 8 (c) −3 ≤ −3
   (d) 0.5 ≥ 0.25 (e) 0 > 9 (f) 0 ≤ 9
   (g) −7 ≥ 0 (h) −7 < 0

1.2   ## Significant figures and decimal places

The accuracy with which we state a number often depends upon the context in which the number is being used. The volume of a petrol tank is usually given to the nearest litre. It is of no practical use to give such a volume to the nearest cubic centimetre.

When writing a number we often give the accuracy by stating the **number of significant figures** or the **number of decimal places** used. These terms are now explained.

Significant figures

Suppose we are asked to write down the number nearest to 6573 using at most 2 non-zero digits. We would write 6600. This number is nearer to 6573 than 6500, and contains only 2 non-zero digits. We say that 6573 to 2 significant figures (2 s.f.) is 6600. To write a number X to 3 s.f. we can use at most 3 non-zero digits and seek the number which is as near as possible to X.

When asked to write a number to, say, 3 s.f. the first step is to look at the first 4 digits. If asked to write the number to 2 s.f. we look at the first 3 digits, and so on. For example, to write 6543.1 to 2 s.f. we would consider the number 6540.0; the digits 3 and 1 are effectively ignored. The next step is to 'round up' or 'round down'. If the final digit is a 5 or more, then we round up by increasing the previous digit by 1. If the final digit is 4 or less, we round down. This leaves the previous digit unchanged although the value of the original number is reduced. Hence to 2 s.f., 6543.1 is 6500. To write 23865 to 3 s.f. we would consider the number 23860. The next step is to increase the 8 to a 9. Thus 23865 is rounded up to 23900. Zeros at the beginning of a number are ignored. To write 0.004693 to 2 s.f. we would firstly consider the number 0.00469. Note that the zeros at the beginning of the number have not been counted. We then round the 6 to a 7, producing 0.0047.

The following examples illustrate the process.

Worked examples

1.3 Write 36.482 to 3 s.f.

Solution We consider the first 4 digits, that is, 36.48. The final digit is 8 and so we round up 36.48 to 36.5. To 3 s.f. 36.482 is 36.5.

1.4 Write 1.0049 to 4 s.f.

Solution To write to 4 s.f. we consider the first 5 digits, that is, 1.0049. The final digit is a 9 and so 1.0049 is rounded up to 1.005.

1.5 Write 695.3 to 2 s.f.

Solution We consider 695. The final digit is a 5 and so we round up. We cannot round up the 9 to a 10 and so the 69 is rounded up to 70. Hence to 2 s.f. the number is 700.

1.6 Write 0.0473 to 1 s.f.

Solution We do not count the initial zeros and consider 0.047. The final digit tells us to round up. Hence to 1 s.f. we have 0.05.

1.7 A number is given to 2 s.f. as 67.
(a) What is the maximum value the number could have?
(b) What is the minimum value the number could have?

Solution (a) To 2 s.f. 67.5 is 68. Any number just below 67.5, for example 67.49 or 67.499, to 2 s.f. is 67. Hence the maximum value of the number is 67.4999....
(b) To 2 s.f. 66.4999... is 66. However, 66.5 to 2 s.f. is 67. The minimum value of the number is thus 66.5.

Decimal places

When asked to write a number to 3 decimal places (3 d.p.) we consider the first 4 decimal places, that is, numbers after the decimal point. If asked to write to 2 d.p. we consider the first 3 decimal places and so on. If the final digit is 5 or more we round up, otherwise we round down.

Worked examples

1.8 Write 63.4261 to 2 d.p.

Solution We consider the number to 3 d.p., that is, 63.426. The final digit is 6 and so we round up 63.426 to 63.43. Hence 63.4261 to 2 d.p. is 63.43.

1.9 Write 1.97 to 1 d.p.

Solution In order to write to 1 d.p. we consider the number to 2 d.p., that is, we consider 1.97. The final digit is a 7 and so we round up. The 9 cannot be rounded up and so we look at 1.9. This can be rounded up to 2.0. Hence 1.97 to 1 d.p. is 2.0. Note that it is crucial to write 2.0 and not simply 2, as this shows that the number is written to 1 d.p.

1.10 Write −6.0439 to 2 d.p.

Solution We consider −6.043. As the final digit is a 3 the number is rounded down to −6.04.

Self-assessment questions 1.2

1. Explain the meaning of 'significant figures'.
2. Explain the process of writing a number to so many decimal places.

Exercise 1.2

1. Write to 3 s.f.

   (a) 6962 (b) 70.406 (c) 0.0123
   (d) 0.010991 (e) 45.607 (f) 2345

2. Write 65.999 to

   (a) 4 s.f. (b) 3 s.f. (c) 2 s.f. (d) 1 s.f.
   (e) 2 d.p. (f) 1 d.p.

3. Write 9.99 to

   (a) 1 s.f. (b) 1 d.p.

4. Write 65.4555 to

   (a) 3 d.p. (b) 2 d.p. (c) 1 d.p. (d) 5 s.f.
   (e) 4 s.f. (f) 3 s.f. (g) 2 s.f. (h) 1 s.f.

1.3   Arithmetical terms and the BODMAS rule

To find the **sum** of two or more numbers, the numbers are added together. To find the **difference** of two numbers, one number is subtracted from the other. The **product** of two numbers is found by multiplying the numbers together. Finally, the **quotient** of two numbers is found by dividing one number by the other.

Worked example

1.11   (a) Find the sum of 3, 6 and 4.
   (b) Find the difference of 6 and 4.
   (c) Find the product of 7 and 2.
   (d) Find the quotient of 9 and 5.

Solution   (a) The sum of 3, 6 and 4 is
       $3 + 6 + 4 = 13$
   (b) The difference of 6 and 4 is
       $6 - 4 = 2$
   (c) The product of 7 and 2 is
       $7 \times 2 = 14$
   (d) The quotient of 9 and 5 is $\frac{9}{5}$.

When writing products we sometimes replace the sign $\times$ by '.' or even omit it completely. For example, $3 \times 6 \times 9$ could be written as 3.6.9 or (3)(6)(9). We note the following important rules for multiplying and dividing positive and negative numbers.

KEY POINT

$$(\text{positive})(\text{positive}) = \text{positive} \qquad \text{and} \qquad \frac{\text{positive}}{\text{positive}} = \text{positive}$$

$$(\text{positive})(\text{negative}) = \text{negative}$$

$$(\text{negative})(\text{positive}) = \text{negative} \qquad\qquad\qquad \frac{\text{positive}}{\text{negative}} = \text{negative}$$

$$(\text{negative})(\text{negative}) = \text{positive}$$

$$\frac{\text{negative}}{\text{positive}} = \text{negative}$$

$$\frac{\text{negative}}{\text{negative}} = \text{positive}$$

Worked example

1.12   Evaluate

(a) $3 \times (-2)$ (b) $(-1) \times 7$ (c) $(-2) \times (-4)$ (d) $\dfrac{12}{(-4)}$ (e) $\dfrac{-8}{4}$ (f) $\dfrac{-6}{-2}$

Solution   (a) $3 \times (-2) = -6$ (b) $(-1) \times 7 = -7$ (c) $(-2) \times (-4) = 8$

(d) $\dfrac{12}{-4} = -3$ (e) $\dfrac{-8}{4} = -2$ (f) $\dfrac{-6}{-2} = 3$

When evaluating numerical expressions we need to know the order in which addition, subtraction, multiplication and division are carried out. As a simple example, consider evaluating $2 + 3 \times 4$. If the addition is carried out first we get $2 + 3 \times 4 = 5 \times 4 = 20$. If the multiplication is carried out first we get $2 + 3 \times 4 = 2 + 12 = 14$. Clearly the order of carrying out numerical operations is important. The BODMAS rule tells us the order in which we must carry out the operations of addition, subtraction, multiplication and division.

KEY POINT

BODMAS stands for

Brackets ( )	First priority
Of $\times$	Second priority
Division $\div$	Second priority
Multiplication $\times$	Second priority
Addition $+$	Third priority
Subtraction $-$	Third priority

This is the order of carrying out arithmetical operations, with bracketed expressions having highest priority and subtraction and addition having the lowest priority. Note that 'Of', 'Division' and 'Multiplication' have equal priority, as do 'Addition' and 'Subtraction'. 'Of' is used to show multiplication when dealing with fractions, for example, find $\frac{1}{2}$ of 6 means $\frac{1}{2} \times 6$.

If an expression contains only multiplication and division, we evaluate by working from left to right. Similarly, if an expression contains only addition and subtraction, we also evaluate by working from left to right.

Worked examples

1.13 Evaluate
(a) $2 + 3 \times 4$ (b) $(2 + 3) \times 4$

Solution (a) Using the BODMAS rule we see that multiplication is carried out first. So
$$2 + 3 \times 4 = 2 + 12 = 14$$
(b) Using the BODMAS rule we see that the bracketed expression takes priority over all else. Hence
$$(2 + 3) \times 4 = 5 \times 4 = 20$$

1.14 Evaluate
(a) $4 - 2 \div 2$ (b) $1 - 3 + 2 \times 2$

Solution (a) Division is carried out before subtraction, and so
$$4 - 2 \div 2 = 4 - \frac{2}{2} = 3$$

Solution (b) Multiplication is carried out before subtraction or addition,
$$1 - 3 + 2 \times 2 = 1 - 3 + 4 = 2$$

1.15 Evaluate
(a) $(12 \div 4) \times 3$ (b) $12 \div (4 \times 3)$

Solution Recall that bracketed expressions are evaluated first.

(a) $(12 \div 4) \times 3 = \left(\dfrac{12}{4}\right) \times 3 = 3 \times 3 = 9$

(b) $12 \div (4 \times 3) = 12 \div 12 = 1$

Example 1.15 shows the importance of the position of brackets in an expression.

Self-assessment questions 1.3

1. State the BODMAS rule used to evaluate expressions.
2. The position of brackets in an expression is unimportant. True or false?
3. Explain the terms sum, difference, product and quotient.

Exercise 1.3

1. Evaluate the following expressions:

   (a) $6 - 2 \times 2$ (b) $(6 - 2) \times 2$
   (c) $6 \div 2 - 2$ (d) $(6 \div 2) - 2$
   (e) $6 - 2 + 3 \times 2$ (f) $6 - (2 + 3) \times 2$
   (g) $(6 - 2) + 3 \times 2$ (h) $\dfrac{16}{-2}$ (i) $\dfrac{-24}{-3}$
   (j) $(-6) \times (-2)$ (k) $(-2)(-3)(-4)$

2. Find the sum and product of (a) 3 and 6, (b) 10 and 7, (c) 2, 3 and 6.

3. Find the difference and quotient of (a) 18 and 9,   (b) 20 and 5, (c) 100 and 20.

4. Place brackets in the following expressions to make them correct:

   (a) $6 \times 12 - 3 + 1 = 55$
   (b) $6 \times 12 - 3 + 1 = 68$
   (c) $6 \times 12 - 3 + 1 = 60$
   (d) $5 \times 4 - 3 + 2 = 7$
   (e) $5 \times 4 - 3 + 2 = 15$
   (f) $5 \times 4 - 3 + 2 = -5$

1.4   Prime numbers and factorization

A **prime number** is a positive integer which cannot be expressed as the product of two smaller positive integers. To put it another way, a prime number is one which can be divided exactly only by 1 and itself. For example, $6 = 2 \times 3$, so 6 can be expressed as a product of smaller numbers and hence 6 is not a prime number. However, 7 is prime. Examples of prime numbers are 1, 2, 3, 5, 7, 11, 13, 17, 19, 23. Note that 2 is the only even prime.

Factorize means 'write as a product'. By writing 12 as 3×4 we have factorized 12. We say 3 is a **factor** of 12 and 4 is also a factor of 12. The way in which a number is factorized is not unique, for example 12 may be expressed as 3×4 or 2×6. Note that 2 and 6 are also factors of 12.

When a number is written as a product of prime numbers we say the number has been **prime factorized**. The prime factorization of a number is unique.

Worked examples

1.16 Prime factorize the following numbers
(a) 12, (b) 40, (c) 70.

Solution (a) We begin with 2 and see if this is a factor of 12. Clearly it is, so we write

$$12 = 2 \times 6$$

Now we consider 6. Again 2 is a factor so we write

$$12 = 2 \times 2 \times 3$$

All the factors are now prime, that is the prime factorization of 12 is $2 \times 2 \times 3$.

(b) Clearly 2 is a factor of 40,

$$40 = 2 \times 20$$

Clearly 2 is a factor of 20,

$$40 = 2 \times 2 \times 10$$

Again 2 is a factor of 10,

$$40 = 2 \times 2 \times 2 \times 5$$

All the factors are now prime. The prime factorization of 40 is $2 \times 2 \times 2 \times 5$.

(c) Clearly 2 is a factor of 70,

$$70 = 2 \times 35$$

We consider 35: 2 is not a factor, 3 is not a factor, but 5 is.

$$70 = 2 \times 5 \times 7$$

All the factors are prime. The prime factorization of 70 is $2 \times 5 \times 7$.

1.17 Prime factorize 2299.

Solution We note that 2 is not a factor and so we try 3. Again 3 is not a factor and so we try 5. This process continues until we find the first prime factor. It is 11,

$$2299 = 11 \times 209$$

We now consider 209. The first prime factor is 11,

$$2299 = 11 \times 11 \times 19$$

All the factors are prime. The prime factorization of 2299 is $11 \times 11 \times 19$.

Self-assessment questions 1.4

1. Explain what is meant by a prime number.

Exercise 1.4

1. Prime factorize the following numbers:

   (a) 26 (b) 100 (c) 27 (d) 71 (e) 64 (f) 87 (g) 437 (h) 899

1.5   Highest common factor and lowest common multiple

Highest common factor

Suppose we prime factorize 12. This gives $12 = 2 \times 2 \times 3$. From this prime factorization we can deduce all the factors of 12.

> 2 is a factor of 12
> 3 is a factor of 12
> $2 \times 2 = 4$ is a factor of 12
> $2 \times 3 = 6$ is a factor of 12

Hence 12 has 4 factors 2, 3, 4 and 6, in addition to the obvious factors of 1 and 12.

Similarly we could prime factorize 18 to obtain $18 = 2 \times 3 \times 3$. From this we can list the factors of 18.

> 2 is a factor of 18
> 3 is a factor of 18
> $2 \times 3 = 6$ is a factor of 18
> $3 \times 3 = 9$ is a factor of 18

The factors of 18 are 1, 2, 3, 6, 9 and 18. Some factors are common to both 12 and 18. These are 2, 3 and 6. These are **common factors** of 12 and 18. The highest common factor of 12 and 18 is 6.

The highest common factor of 12 and 18 can be obtained directly from their prime factorization. We simply note all the primes common to both factorizations:

$$12 = 2 \times 2 \times 3 \qquad 18 = 2 \times 3 \times 3$$

Common to both is 2×3. Thus the highest common factor is $2 \times 3 = 6$. Thus, 6 is the highest number which divides both 12 and 18.

KEY POINT

> Given 2 or more numbers the **highest common factor** (h.c.f.) is the largest (highest) number which is a factor of all the given numbers.

Worked examples

1.18   Find the h.c.f. of 12 and 27.

Solution   We prime factorize 12 and 27,

$$12 = 2 \times 2 \times 3 \qquad 27 = 3 \times 3 \times 3$$

Common to both is 3. Thus, 3 is the h.c.f. of 12 and 27. This means that 3 is the highest number which divides both 12 and 27.

1.19   Find the h.c.f. of 28 and 210.

Solution   The numbers are prime factorized,

$$28 = 2 \times 2 \times 7$$

$$210 = 2 \times 3 \times 5 \times 7$$

The factors which are common are identified: a 2 is common to both and a 7 is common to both. Hence both numbers are divisible by $2 \times 7 = 14$. Since this number contains all the common factors it is the highest common factor.

1.20   Find the the h.c.f. of 90 and 108.

Solution   The numbers are prime factorized,

$$90 = 2 \times 3 \times 3 \times 5$$

$$108 = 2 \times 2 \times 3 \times 3 \times 3$$

The common factors are 2, 3 and 3 and so the h.c.f. is $2 \times 3 \times 3$, that is 18. This is the highest number which divides both 90 and 108.

1.21   Find the h.c.f. of 12, 18 and 20.

Solution   Prime factorization yields

$$12 = 2 \times 2 \times 3 \qquad 18 = 2 \times 3 \times 3 \qquad 20 = 2 \times 2 \times 5$$

There is only one factor common to all 3 numbers; it is 2. Hence 2 is the h.c.f. of 12, 18 and 20.

Lowest common multiple

Suppose we are given 2 or more numbers and wish to find numbers into which all the given numbers will divide. For example, given 4 and 6 we see that they both divide exactly into 12, 24, 36, 48, 60 and so on. The smallest number into which they both divide is 12. We say 12 is the **lowest common multiple** of 4 and 6.

KEY POINT

> The lowest common multiple (l.c.m.) of a set of numbers is the smallest (lowest) number into which all the given numbers will divide exactly.

Worked example

1.22   Find the l.c.m. of 6 and 10.

Solution   We seek the smallest number into which both 6 and 10 will divide exactly. There are many numbers into which 6 and 10 will divide, for example 60, 120, 600, but we are seeking the smallest such number. By inspection, the smallest such number is 30. Thus, the l.c.m. of 6 and 10 is 30.

A more systematic method of finding the l.c.m. involves the use of prime factorization.

Worked examples

1.23   Find the l.c.m. of 15 and 20.

Solution   The numbers are prime factorized,

$$15 = 3 \times 5 \qquad 20 = 2 \times 2 \times 5$$

We require the smallest number into which both of the given numbers will divide. Consider the number of 2s required in the l.c.m. The highest number of 2s occurs in the factorization of 20. Hence the l.c.m. requires 2 factors of 2. Consider the number of 3s required. The highest number of 3s occurs in the factorization of 15. Hence the l.c.m. requires 1 factor of 3. Consider the number of 5s required. The highest number of 5s is 1 and so the l.c.m. requires 1 factor of 5. Hence the l.c.m. is $2 \times 2 \times 3 \times 5 = 60$.

1.24   Find the l.c.m. of 20, 24 and 25.

Solution   The numbers are prime factorized,

$$20 = 2 \times 2 \times 5 \quad 24 = 2 \times 2 \times 2 \times 3 \quad 25 = 5 \times 5$$

Consider the number of 2s required. The highest number of 2s required is 3 from factorizing 24. The highest number of 3s required is 1, again from factorizing 24. The highest number of 5s required is 2, found from factorizing 25. Hence the l.c.m. is given by

$$\text{l.c.m.} = 2 \times 2 \times 2 \times 3 \times 5 \times 5 = 600$$

Hence 600 is the smallest number into which 20, 24 and 25 will all divide exactly.

Self-assessment questions 1.5

1. Explain what is meant by the h.c.f. of a set of numbers.
2. Explain what is meant by the l.c.m. of a set of numbers.

Exercise 1.5

1. Calculate the h.c.f. of the following sets of numbers:
   (a) 12, 15, 21 (b) 16, 24, 40 (c) 28, 70, 120, 160 (d) 35, 38, 42 (e) 96, 120, 144

2. Calculate the l.c.m. of the following sets of numbers:
   (a) 5, 6, 8 (b) 20, 30 (c) 7, 9, 12 (d) 100, 150, 235 (e) 96, 120, 144

Test and assignment exercises 1

1. Evaluate
   (a) $6 \div 2 + 1$ (b) $6 \div (2 + 1)$
   (c) $12 + 4 \div 4$ (d) $(12 + 4) \div 4$
   (e) $3 \times 2 + 1$ (f) $3 \times (2 + 1)$
   (g) $6 - 2 + 4 \div 2$ (h) $(6 - 2 + 4) \div 2$
   (i) $6 - (2 + 4 \div 2)$ (j) $6 - (2 + 4) \div 2$
   (k) $2 \times 4 - 1$ (l) $2 \times (4 - 1)$
   (m) $2 \times 6 \div (3 - 1)$ (n) $2 \times (6 \div 3) - 1$
   (o) $2 \times (6 \div 3 - 1)$

2. Prime factorize (a) 56, (b) 39, (c) 74.

3. Find the h.c.f. of
   (a) 8, 12, 14
   (b) 18, 42, 66
   (c) 20, 24, 30
   (d) 16, 24, 32, 160

4. Find the l.c.m of
   (a) 10, 15
   (b) 11, 13
   (c) 8, 14, 16
   (d) 15, 24, 30

5. Write 0.09846 to (a) 1 d.p., (b) 2 s.f., (c) 1 s.f.

6. Write 9.513 to (a) 3 s.f., (b) 2 s.f., (c) 1 s.f.

7. Write 19.96 to (a) 1 d.p., (b) 2 s.f., (c) 1 s.f.

2 Fractions

Objectives	This chapter
	• explains how to add, subtract, multiply and divide fractions
	• defines the terms 'improper fraction', 'proper fraction' and 'mixed fraction'

A fraction is a number of the form p/q where p and q are integers with $q \neq 0$. The **numerator** is p and the **denominator** is q. If $p \geq q$, the fraction is **improper**. If $p < q$ then the fraction is **proper**. For example, $\frac{7}{4}$ and $\frac{3}{3}$ are improper fractions and $\frac{7}{9}$ and $\frac{11}{15}$ are proper fractions.

2.1   Improper fractions

Consider the improper fraction $\frac{11}{4}$. Recognizing that 2 is $\frac{8}{4}$ we may write $\frac{11}{4}$ as

$$\frac{11}{4} = \frac{8+3}{4} = \frac{8}{4} + \frac{3}{4} = 2 + \frac{3}{4} = 2\frac{3}{4}$$

The number $2\frac{3}{4}$ is referred to as a **mixed fraction**. Alternatively we can think of 4 dividing into 11 twice with a remainder of 3.

Worked examples

2.1   Express as mixed fractions: (a) $\frac{19}{3}$ (b) $\frac{25}{7}$

Solution

(a) $\dfrac{19}{3} = \dfrac{18+1}{3} = \dfrac{18}{3} + \dfrac{1}{3} = 6 + \dfrac{1}{3} = 6\dfrac{1}{3}$

(b) $\dfrac{25}{7} = \dfrac{21+4}{7} = \dfrac{21}{7} + \dfrac{4}{7} = 3 + \dfrac{4}{7} = 3\dfrac{4}{7}$

2.2 Express $4\frac{1}{5}$ as an improper fraction.

Solution

$$4\frac{1}{5} = 4 + \frac{1}{5} = \frac{20}{5} + \frac{1}{5} = \frac{21}{5}$$

So $4\frac{1}{5}$ is identical to $\frac{21}{5}$.

Self-assessment questions 2.1

1. Explain the terms 'proper fraction' and 'improper fraction'.
2. All integers can be thought of as fractions. True or false?

Exercise 2.1

1. Express as mixed fractions:

   (a) $\frac{5}{3}$ (b) $\frac{9}{4}$ (c) $\frac{7}{5}$ (d) $-\frac{20}{11}$ (e) $\frac{11}{5}$

2. Express as improper fractions:

   (a) $2\frac{1}{2}$ (b) $3\frac{2}{3}$ (c) $10\frac{1}{4}$ (d) $5\frac{2}{7}$ (e) $6\frac{2}{9}$

2.2   **Expressing a fraction in its simplest form**

Given a fraction, we may be able to express it in a simpler form. We note that multiplying or dividing both numerator and denominator by the same number leaves the value of the fraction unchanged. For example, given $\frac{6}{12}$, we can divide both numerator and denominator by 2 to obtain $\frac{3}{6}$, and these two fractions have the same value. We say that $\frac{3}{6}$ and $\frac{6}{12}$ are **equivalent fractions**. Similarly, given $\frac{5}{12}$ we can multiply both numerator and denominator by say 4 to obtain $\frac{20}{48}$, and these two fractions have the same value, that is they are equivalent.

A fraction is in its simplest form when there are no factors common to both numerator and denominator. For example, $\frac{5}{12}$ is in its simplest form, but $\frac{3}{6}$ is not since 3 is a factor common to both numerator and denominator. In its simplest form $\frac{3}{6}$ is $\frac{1}{2}$.

Worked examples

2.3 Express $\frac{24}{36}$ in its simplest form.

Solution We seek factors common to both numerator and denominator. To do this we prime factorize 24 and 36.

$$24 = 2 \times 2 \times 2 \times 3 \qquad 36 = 2 \times 2 \times 3 \times 3$$

Prime factorization has been described in §1.4

The factors $2 \times 2 \times 3$ are common to both 24 and 36 and so these may be cancelled. Note that only factors may be cancelled when simplifying

Finding the highest common factor (h.c.f.) of two numbers is detailed in §1.5

a fraction. Hence

$$\frac{24}{36} = \frac{\cancel{2} \times \cancel{2} \times 2 \times \cancel{3}}{\cancel{2} \times \cancel{2} \times \cancel{3} \times 3} = \frac{2}{3}$$

In its simplest form $\frac{24}{36}$ is $\frac{2}{3}$. In effect we have divided 24 and 36 by 12, which is their h.c.f.

2.4   Express $\frac{49}{21}$ in its simplest form.

Solution   Prime factorizing 49 and 21 gives

$$49 = 7 \times 7 \qquad 21 = 3 \times 7$$

Their h.c.f. is 7. Dividing 49 and 21 by 7 gives

$$\frac{49}{21} = \frac{7}{3}$$

Hence the simplest form of $\frac{49}{21}$ is $\frac{7}{3}$.

Self-assessment questions 2.2

1. Explain the use of h.c.f. in the simplification of fractions.
2. Give an example of 3 fractions which are equivalent.

Exercise 2.2

1. Express the following fractions in their simplest form:

(a) $\frac{18}{27}$ (b) $\frac{12}{20}$ (c) $\frac{15}{45}$

(d) $\frac{25}{80}$ (e) $\frac{15}{60}$ (f) $\frac{90}{200}$

(g) $\frac{15}{20}$ (h) $\frac{2}{18}$ (i) $\frac{16}{24}$

(j) $\frac{30}{65}$ (k) $\frac{12}{21}$ (l) $\frac{100}{45}$

2.3   Addition and subtraction of fractions

To add and subtract fractions we first express them all with the same denominator. This common denominator is the l.c.m. of the original denominators.

Worked examples

2.5   Find $\frac{2}{3} + \frac{5}{4}$.

Solution   The denominators are 3 and 4. The l.c.m. of 3 and 4 is 12. We need to express both fractions with a denominator of 12.

Finding the lowest common multiple (l.c.m.) is detailed in §1.5

To express $\frac{2}{3}$ with a denominator of 12 we multiply both numerator and denominator by 4. Hence $\frac{2}{3}$ is the same as $\frac{8}{12}$. To express $\frac{5}{4}$ with a denominator of 12 we multiply both numerator and denominator by 3. Hence $\frac{5}{4}$ is the same as $\frac{15}{12}$. So

$$\frac{2}{3} + \frac{5}{4} = \frac{8}{12} + \frac{15}{12} = \frac{8+15}{12} = \frac{23}{12}$$

2.6 Find $\frac{4}{9} - \frac{1}{2} + \frac{5}{6}$.

Solution The denominators are 9, 2 and 6. Their l.c.m. is 18. Each fraction is expressed with 18 as the denominator

$$\frac{4}{9} = \frac{8}{18} \qquad \frac{1}{2} = \frac{9}{18} \qquad \frac{5}{6} = \frac{15}{18}$$

Then

$$\frac{4}{9} - \frac{1}{2} + \frac{5}{6} = \frac{8}{18} - \frac{9}{18} + \frac{15}{18} = \frac{8-9+15}{18} = \frac{14}{18}$$

The fraction $\frac{14}{18}$ can be simplified to $\frac{7}{9}$. Hence

$$\frac{4}{9} - \frac{1}{2} + \frac{5}{6} = \frac{7}{9}$$

Self-assessment questions 2.3

1. Explain the use of l.c.m. when adding and subtracting fractions.

Exercise 2.3

1. Find

(a) $\frac{1}{4} + \frac{2}{3}$ (b) $\frac{3}{5} + \frac{5}{3}$ (c) $\frac{12}{14} - \frac{2}{7}$ (f) $2\frac{1}{4} - 1\frac{1}{3} + \frac{1}{2}$ (g) $\frac{10}{15} - 1\frac{2}{5} + \frac{8}{3}$

(d) $\frac{3}{7} - \frac{1}{2} + \frac{2}{21}$ (e) $1\frac{1}{2} + \frac{4}{9}$ (h) $\frac{9}{10} - \frac{7}{16} + \frac{1}{2} - \frac{2}{5}$

2.4   **Multiplication of fractions**

The product of 2 or more fractions is found by multiplying their numerators to form a new numerator, and then multiplying their denominators to form a new denominator.

Worked examples

2.7 Find $\frac{4}{9} \times \frac{3}{8}$.

Solution The numerators are multiplied: $4 \times 3 = 12$. The denominators are multiplied: $9 \times 8 = 72$. Hence

$$\frac{4}{9} \times \frac{3}{8} = \frac{12}{72}$$

This may now be expressed in its simplest form,

$$\frac{12}{72} = \frac{1}{6}$$

Hence

$$\frac{4}{9} \times \frac{3}{8} = \frac{1}{6}$$

An alternative, but equivalent, method is to cancel any factors common to both numerator and denominator at the outset:

$$\frac{4}{9} \times \frac{3}{8} = \frac{4 \times 3}{9 \times 8}$$

A factor of 4 is common to the 4 and the 8. Hence

$$\frac{4 \times 3}{9 \times 8} = \frac{1 \times 3}{9 \times 2}$$

A factor of 3 is common to the 3 and the 9. Hence

$$\frac{1 \times 3}{9 \times 2} = \frac{1 \times 1}{3 \times 2} = \frac{1}{6}$$

2.8 Find $\frac{12}{25} \times \frac{2}{7} \times \frac{10}{9}$.

Solution We cancel factors common to both numerator and denominator. A factor of 5 is common to 10 and 25. Cancelling this gives

$$\frac{12}{25} \times \frac{2}{7} \times \frac{10}{9} = \frac{12}{5} \times \frac{2}{7} \times \frac{2}{9}$$

A factor of 3 is common to 12 and 9. Cancelling this gives

$$\frac{12}{5} \times \frac{2}{7} \times \frac{2}{9} = \frac{4}{5} \times \frac{2}{7} \times \frac{2}{3}$$

There are no more common factors. Hence

$$\frac{12}{25} \times \frac{2}{7} \times \frac{10}{9} = \frac{4}{5} \times \frac{2}{7} \times \frac{2}{3} = \frac{16}{105}$$

2.9 Find $\frac{3}{4}$ of $\frac{5}{9}$.

Solution $\frac{3}{4}$ of $\frac{5}{9}$ is the same as $\frac{3}{4} \times \frac{5}{9}$. Cancelling a factor of 3 from numerator and denominator gives $\frac{1}{4} \times \frac{5}{3}$, that is $\frac{5}{12}$. Hence $\frac{3}{4}$ of $\frac{5}{9}$ is $\frac{5}{12}$.

Recall that 'of' means multiply

2.10 Find $\frac{5}{6}$ of 70.

Solution We can write 70 as $\frac{70}{1}$. So

$$\frac{5}{6} \text{ of } 70 = \frac{5}{6} \times \frac{70}{1} = \frac{5}{3} \times \frac{35}{1} = \frac{175}{3} = 58\frac{1}{3}$$

Self-assessment questions 2.4

1. Describe how to multiply fractions together.

Exercise 2.4

1. Evaluate

(a) $\frac{2}{3} \times \frac{6}{7}$ (b) $\frac{8}{15} \times \frac{25}{32}$ (c) $\frac{1}{4} \times \frac{8}{9}$

(d) $\frac{16}{17} \times \frac{34}{48}$ (e) $2 \times \frac{3}{5} \times \frac{5}{12}$ (f) $2\frac{1}{3} \times 1\frac{1}{4}$

(g) $1\frac{3}{4} \times 2\frac{1}{2}$ (h) $\frac{3}{4} \times 1\frac{1}{2} \times 3\frac{1}{2}$

2. Evaluate

(a) $\frac{2}{3}$ of $\frac{3}{4}$ (b) $\frac{4}{7}$ of $\frac{21}{30}$ (c) $\frac{9}{10}$ of 80

(d) $\frac{6}{7}$ of 42

3. Is $\frac{3}{4}$ of $\frac{12}{15}$ the same as $\frac{12}{15}$ of $\frac{3}{4}$?

2.5   Division by a fraction

When dividing by a fraction, the fraction is inverted and then multiplied. This process is illustrated and explained in the following example.

Worked examples

2.11 Find $\frac{6}{25} \div \frac{2}{5}$.

Solution We invert $\frac{2}{5}$ to obtain $\frac{5}{2}$. Multiplication is then performed. So

$$\frac{6}{25} \div \frac{2}{5} = \frac{6}{25} \times \frac{5}{2} = \frac{3}{25} \times \frac{5}{1} = \frac{3}{5} \times \frac{1}{1} = \frac{3}{5}$$

2.12 Evaluate (a) $1\frac{1}{3} \div \frac{8}{3}$ (b) $\frac{20}{21} \div \frac{5}{7}$.

Solution   (a) Firstly we express $1\frac{1}{3}$ as an improper fraction.

$$1\frac{1}{3} = 1 + \frac{1}{3} = \frac{3}{3} + \frac{1}{3} = \frac{4}{3}$$

So we calculate $\dfrac{4}{3} \div \dfrac{8}{3} = \dfrac{4}{3} \times \dfrac{3}{8} = \dfrac{4}{8} = \dfrac{1}{2}$

Hence $1\dfrac{1}{3} \div \dfrac{8}{3} = \dfrac{1}{2}$

(b) $\dfrac{20}{21} \div \dfrac{5}{7} = \dfrac{20}{21} \times \dfrac{7}{5} = \dfrac{4}{21} \times \dfrac{7}{1} = \dfrac{4}{3}$

Self-assessment questions 2.5

1. Explain the process of division by a fraction.

Exercise 2.5

1. Evaluate

(a) $\dfrac{3}{4} \div \dfrac{1}{8}$ (b) $\dfrac{8}{9} \div \dfrac{4}{3}$ (c) $\dfrac{-2}{7} \div \dfrac{4}{21}$

(d) $\dfrac{9}{4} \div 1\dfrac{1}{2}$ (e) $\dfrac{5}{6} \div \dfrac{5}{12}$ (f) $\dfrac{99}{100} \div 1\dfrac{4}{5}$

(g) $3\dfrac{1}{4} \div 1\dfrac{1}{8}$ (h) $\left(2\dfrac{1}{4} \div \dfrac{3}{4}\right) \times 2$

(i) $2\dfrac{1}{4} \div \left(\dfrac{3}{4} \times 2\right)$ (j) $6\dfrac{1}{4} \div 2\dfrac{1}{2} + 5$

(k) $6\dfrac{1}{4} \div \left(2\dfrac{1}{2} + 5\right)$

Test and assignment exercises 2

1. Evaluate

(a) $\dfrac{3}{4} + \dfrac{1}{6}$ (b) $\dfrac{2}{3} + \dfrac{3}{5} - \dfrac{1}{6}$

(c) $\dfrac{5}{7} - \dfrac{2}{3}$ (d) $2\dfrac{1}{3} - \dfrac{9}{10}$

(e) $5\dfrac{1}{4} + 3\dfrac{1}{6}$ (f) $\dfrac{9}{8} - \dfrac{7}{6} + 1$

(g) $\dfrac{5}{6} - \dfrac{5}{3} + \dfrac{5}{4}$ (h) $\dfrac{4}{5} + \dfrac{1}{3} - \dfrac{3}{4}$

2. Evaluate

(a) $\dfrac{4}{7} \times \dfrac{21}{32}$ (b) $\dfrac{5}{6} \times \dfrac{8}{15}$

(c) $\dfrac{3}{11} \times \dfrac{20}{21}$ (d) $\dfrac{9}{14} \times \dfrac{8}{18}$

(e) $\dfrac{5}{4} \div \dfrac{10}{13}$ (f) $\dfrac{7}{16} \div \dfrac{21}{32}$

(g) $\dfrac{-24}{25} \div \dfrac{51}{50}$ (h) $\dfrac{45}{81} \div \dfrac{25}{27}$

3. Evaluate the following expressions using the BODMAS rule.

(a) $\dfrac{1}{2} + \dfrac{1}{3} \times 2$ (b) $\dfrac{3}{4} \times \dfrac{2}{3} + \dfrac{1}{4}$

(c) $\dfrac{5}{6} \div \dfrac{2}{3} + \dfrac{3}{4}$ (d) $\left(\dfrac{2}{3} + \dfrac{1}{4}\right) \div 4 + \dfrac{3}{5}$

(e) $\left(\dfrac{4}{3} - \dfrac{2}{5} \times \dfrac{1}{3}\right) \times \dfrac{1}{4} + \dfrac{1}{2}$

(f) $\dfrac{3}{4}$ of $\left(1 + \dfrac{2}{3}\right)$

(g) $\dfrac{2}{3}$ of $\dfrac{1}{2} + 1$ (h) $\dfrac{1}{5} \times \dfrac{2}{3} + \dfrac{2}{5} \div \dfrac{4}{5}$

4. Express in their simplest form:

(a) $\dfrac{21}{84}$ (b) $\dfrac{6}{80}$ (c) $\dfrac{34}{85}$ (d) $\dfrac{22}{143}$ (e) $\dfrac{69}{253}$

3 Percentage and ratio

Objectives

This chapter

- explains the terms 'percentage' and 'ratio'
- shows how to perform calculations using percentages and ratios

3.1 Percentage

A **percentage** is a fraction whose denominator is 100. We use the per-cent symbol % to represent a percentage, for example $\frac{17}{100}$ may be written as 17%.

Worked example

3.1 Express $\frac{2}{5}$ as a percentage.

Solution We need to express $\frac{2}{5}$ as a fraction with a denominator of 100. To do this we multiply the denominator by 100. In order not to alter the value of the fraction we must also multiply the numerator by 100. Thus

$$\frac{2}{5} = \frac{2}{5} \times \frac{100}{100}$$

We may write this last expression as

$$\frac{2}{5} \times \frac{100}{100} = \left(\frac{2}{5} \times 100\right) \times \frac{1}{100}$$

$$= \left(\frac{200}{5}\right) \times \frac{1}{100}$$

$$= 40 \times \frac{1}{100}$$

$$= \frac{40}{100}$$

$$= 40\%$$

Hence $\frac{2}{5}$ is equivalent to 40%.

As an easy workable rule we can convert a fraction to a percentage by multiplying the fraction by 100, and then labelling the result as a percentage. For example

$$\frac{2}{5} \times 100 = 40$$

Hence $\frac{2}{5}$ as a percentage is 40%.

Worked examples

3.2 Bill scores $\frac{13}{17}$ in a test. Mary scores $\frac{14}{19}$. Express the scores as percentages.

Solution

$$\frac{13}{17} = \frac{13}{17} \times \frac{100}{100} = \left(\frac{13 \times 100}{17}\right) \times \frac{1}{100} = \frac{76.5}{100} = 76.5\%$$

$$\frac{14}{19} = \frac{14}{19} \times \frac{100}{100} = \left(\frac{14 \times 100}{19}\right) \times \frac{1}{100} = \frac{73.7}{100} = 73.7\%$$

In this form it is easier to make comparisons of the two marks.

3.3 Calculate 27% of 90.

Solution Recall that 'of' means multiply.

$$27\% \text{ of } 90 = \frac{27}{100} \times 90 = \frac{27}{10} \times 9 = 24.3$$

From the example 3.4 we note that 100% of a number is simply the number itself.

3.4 Calculate 100% of 6.

Solution

$$100\% \text{ of } 6 = \frac{100}{100} \times 6 = 1 \times 6 = 6$$

3.5 A deposit of £750 increases by 9% . Calculate the resulting deposit.

Solution We calculate 9% of 750.

$$9\% \text{ of } 750 = \frac{9}{100} \times 750 = 67.5$$

The deposit has increased by £67.5. The resulting deposit is 750 + 67.5 = £817.5.

Alternatively we may perform the calculation as follows. The original deposit represents 100%. The deposit increases by 9% to 109% of the original. So the resulting deposit is 109% of £750.

$$109\% \text{ of } 750 = \frac{109}{100} \times 750 = 817.5$$

3.6   Express 30% as a decimal.

Solution

$$30\% = \frac{30}{100} = 0.3$$

Hence 30% is equivalent to 0.3.

3.7   A television set is advertised at £315. The retailer offers a 10% discount. How much do you pay for the television?

Solution

$$10\% \text{ of } 315 = \frac{10}{100} \times 315 = 31.5$$

The discount is £31.5 and so the cost is 315 − 31.5 = £283.5.

Self-assessment questions 3.1

1. Give one reason why it is sometimes useful to express fractions as percentages.

Exercise 3.1

1. Calculate 23% of 124.

2. Express the following as percentages:

   (a) $\dfrac{9}{11}$ (b) $\dfrac{15}{20}$ (c) $\dfrac{9}{10}$ (d) $\dfrac{45}{50}$ (e) $\dfrac{75}{90}$

3. Express $\dfrac{13}{12}$ as a percentage.

4. Calculate 217% of 500.

5. A worker earns £400 a week. She receives a 6% increase. Calculate her new weekly wage.

6. A debt of £1200 is decreased by 17%. Calculate the remaining debt.

7. Express the following percentages as decimals:

   (a) 50% (b) 36% (c) 75% (d) 100% (e) 12.5%

3.2   Ratio

Ratios are simply an alternative way of expressing fractions. Consider the problem of dividing £200 between 2 people, Ann and Bill, in the ratio of 7:3. This means that Ann receives £7 for every £3 that Bill receives. So, every £10 is divided as £7 to Ann and £3 to Bill. Thus Ann receives $\frac{7}{10}$ of the money. Now $\frac{7}{10}$ of £200 is $\frac{7}{10} \times 200 = £140$. So Ann receives £140 and Bill receives £60.

Worked examples

3.8   Divide 170 in the ratio 3:2.

Solution   A ratio of 3:2 means that every 5 units are split as 3 and 2, that is, the first number is $\frac{3}{5}$ of the total; the second number is $\frac{2}{5}$ of the total.

$$\frac{3}{5} \text{ of } 170 = \frac{3}{5} \times 170 = 102$$

$$\frac{2}{5} \text{ of } 170 = \frac{2}{5} \times 170 = 68$$

The number is divided into 102 and 68.

3.9   Divide 250 cm in the ratio 1:3:4.

Solution   Every 8 cm is divided into 1 cm, 3 cm and 4 cm. Thus, the first length is $\frac{1}{8}$ of the total, the second length is $\frac{3}{8}$ of the total and the third length is $\frac{4}{8}$ of the total.

$$\frac{1}{8} \text{ of } 250 = \frac{1}{8} \times 250 = 31.25$$

$$\frac{3}{8} \text{ of } 250 = \frac{3}{8} \times 250 = 93.75$$

$$\frac{4}{8} \text{ of } 250 = \frac{4}{8} \times 250 = 125$$

The 250 cm length is divided into 31.25 cm, 93.75 cm and 125 cm.

Self-assessment questions 3.2

1. Dividing a number in the ratio 2:3 is the same as dividing it in the ratio of 10:15. True or false?

Exercise 3.2

1. Divide 180 in the ratio 8:1:3.

2. Divide 930 cm in the ratio 1:1:3.

3. A 6 m length of wood is cut in the ratio 2:3:4. Calculate the length of each piece.

4. Divide 1200 in the ratio 1:2:3:4.

5. A sum of £2600 is divided between Alan, Bill and Claire in the ratio of $2\frac{3}{4}:1\frac{1}{2}:2\frac{1}{4}$. Calculate the amount that each receives.

Test and assignment exercises 3

1. Express as decimals
   (a) 8% (b) 18% (c) 65%

2. Express as percentages
   (a) $\frac{3}{8}$ (b) $\frac{79}{100}$ (c) $\frac{56}{118}$

3. Calculate 27.3% of 1496.

4. Calculate 125% of 125.

5. Calculate 85% of 0.25.

6. Divide 0.5 in the ratio of 2:4:9

7. A bill totals £234.5 to which is added tax at 17.5%. Calculate the amount of tax to be paid.

8. An inheritance is divided between 3 people in the ratio 4:7:2. If the least amount received is £2300 calculate how much the other two people received.

9. Divide 70 in the ratio of 0.5:1.3:2.1.

10. Divide 50% in the ratio 2:3.

4 Algebra

Objectives

This chapter

- explains what is meant by 'algebra'
- introduces important algebraic notations
- explains what is meant by a 'power' or 'index'
- explains what is meant by a 'formula'

4.1   What is algebra?

In order to extend the techniques of arithmetic so that they can be more useful in applications we introduce letters or **symbols** to represent quantities of interest. For example, we may choose the capital letter I to stand for the *interest rate* in a business calculation, or the lower case letter t to stand for the *time* in a scientific calculation, and so on. The choice of which letter to use for which quantity is largely up to the user although some conventions have been developed. Very often the letters x and y are used to stand for arbitrary quantities. **Algebra** is the body of mathematical knowledge which has been developed to manipulate symbols. Some symbols take fixed and unchanging values, and these are known as **constants.** For example, suppose we let the symbol b stand for the boiling point of water. This is fixed at 100 °C and so b is a constant. Some symbols represent quantities which can vary, and these are called **variables.** For example, the velocity of a car might be represented by the symbol v, and might vary from 0 to 100 kilometres per hour.

Algebraic notation

In algebraic work particular attention must be paid to the type of symbol used, so that, for example, the symbol T is quite different from the

Table 4.1
The Greek alphabet

A	α	alpha	I	ι	iota	P	ρ	rho
B	β	beta	K	κ	kappa	Σ	σ	sigma
Γ	γ	gamma	Λ	λ	lambda	T	τ	tau
Δ	δ	delta	M	μ	mu	Y	υ	upsilon
E	ϵ	epsilon	N	ν	nu	Φ	ϕ	phi
Z	ζ	zeta	Ξ	ξ	xi	X	χ	chi
H	η	eta	O	o	omicron	Ψ	ψ	psi
Θ	θ	theta	Π	π	pi	Ω	ω	omega

Your scientific calculator is pre-programmed with the value of π. Check that you can use it

symbol t. Usually the symbols chosen are letters from the English alphabet although we frequently meet Greek letters. You may already be aware that the Greek letter 'pi' which has the symbol π is used in the formula for the area of a circle, and is equal to the constant 3.14159.... In many calculations π can be approximated by $\frac{22}{7}$. For reference the full Greek alphabet is given in Table 4.1.

Another important feature is the position of a symbol in relation to other symbols. As we shall see in this chapter, the quantities xy, x^y, y^x and x_y all can mean quite different things. When a symbol is placed to the right and slightly higher than another symbol it is referred to as a **super-script**. So, the quantity x^y contains the superscript y. Likewise, if a symbol is placed to the right and slightly lower than another symbol it is called a **subscript**. The quantity x_1 contains the subscript 1.

The arithmetic of symbols

Addition $(+)$ If the letters x and y stand for two numbers, their **sum** is written as $x + y$. Note that $x + y$ is the same as $y + x$ just as $4 + 7$ is the same as $7 + 4$.

Subtraction $(-)$ The quantity $x - y$ is called the **difference** of x and y, and means the number y subtracted from the number x. Note that $x - y$ is not the same as $y - x$, in the same way that $5 - 3$ is different from $3 - 5$.

Multiplication $(\times)$ Five times the number x is written $5 \times x$, although when multiplying the $\times$ sign is sometimes replaced with '.', or is even left out altogether. This means that $5 \times x$, $5.x$ and $5x$ all mean five times the number x. Similarly $x \times y$ can be written $x.y$ or simply xy. When multiplying, the order of the symbols is not important so that xy

is the same as yx just as 5×4 is the same as 4×5. xy is also known as the **product** of x and y.

Division ($\div$) $x \div y$ means the number x divided by the number y. This is also written x/y. Here the order is important and x/y is quite different from y/x. An expression involving one symbol divided by another is known as an **algebraic fraction.** The top line is called the **numerator** and the bottom line is called the **denominator.** x/y is known as the **quotient** of x and y.

A quantity made up of symbols together with $+$, $-$, $\times$ or $\div$ is called an **algebraic expression.** When evaluating an algebraic expression the BOD-MAS rule given in Chapter 1 applies. This rule reminds us of the correct order in which to evaluate an expression.

Self-assessment questions 4.1

1. Explain what you understand by the term 'algebra'.
2. If m and n are two numbers, explain what is meant by mn.
3. What is an algebraic fraction? Explain the meaning of the terms 'numerator' and 'denominator'.
4. What is the distinction between a superscript and a subscript?
5. What is the distinction between a variable and a constant?

4.2   Powers, or indices

Frequently we shall need to multiply a number by itself several times. For example $3 \times 3 \times 3$, or $a \times a \times a \times a$.

To abbreviate such quantities a new notation is introduced. $a \times a \times a$ is written a^3, pronounced 'a cubed'. The superscript 3 is called a **power** or **index** and the letter a is called the **base.** Similarly $a \times a$ is written a^2, pronounced 'a squared' or 'a raised to the power 2'.

The power is the number of times that the base is multiplied by itself. Most calculators have a button marked x^y which can be used to evaluate expressions such as 2^8, 3^{11} and so on. Check to see if your calculator can do these by verifying that $2^8 = 256$ and $3^{11} = 177147$. Note that the plural of index is **indices.**

As a^2 means $a \times a$, and a^3 means $a \times a \times a$, then we interpret a^1 as simply a. That is, any number raised to the power 1 is itself.

The calculator button x^y is used to find powers of numbers

KEY POINT   Any number raised to the power 1 is itself, that is $a^1 = a$.

Worked examples

4.1 In the expression 3^8 identify the index and the base.

Solution In the expression 3^8, the index is 8 and the base is 3.

4.2 Explain what is meant by y^5.

Solution y^5 means $y \times y \times y \times y \times y$.

4.3 Explain what is meant by $x^2 y^3$.

Solution x^2 means $x \times x$; y^3 means $y \times y \times y$. Therefore $x^2 y^3$ means $x \times x \times y \times y \times y$.

4.4 Evaluate 2^3 and 3^4.

Solution 2^3 means $2 \times 2 \times 2$, that is, 8. Similarly 3^4 means $3 \times 3 \times 3 \times 3$, that is 81.

4.5 Explain what is meant by 7^1.

Solution Any number to the power 1 is itself, that is 7^1 is simply 7.

4.6 Evaluate 10^2 and 10^3.

Solution 10^2 means 10×10 or 100. Similarly 10^3 means $10 \times 10 \times 10$ or 1000.

4.7 Use indices to write the expression $a \times a \times b \times b \times b$ more compactly.

Solution $a \times a$ can be written a^2; $b \times b \times b$ can be written b^3. Therefore $a \times a \times b \times b \times b$ can be written as $a^2 \times b^3$ or simply $a^2 b^3$.

4.8 Write out fully $z^3 y^2$.

Solution $z^3 y^2$ means $z \times z \times z \times y \times y$. Note we could also write this as $zzzyy$.

In an expression such as $3x^2$ care must be taken to ensure the correct use of the power. $3x^2$ means $3 \times x \times x$, so that the power 2 only applies to x. Brackets can significantly change the meaning of an expression. For example, $(3x)^2$ means $(3x) \times (3x)$, that is, $9x^2$. Clearly, $(3x)^2$ is quite different from $3x^2$.

Worked examples

4.9 Explain the distinction between ab^2 and $(ab)^2$.

Solution ab^2 means $a \times b \times b$ whereas $(ab)^2$ means $(ab) \times (ab)$ which equals $a \times b \times a \times b$. This could also be written as $a^2 b^2$.

4.10   Explain the meanings of $-x^2$ and $(-x)^2$. Are these different?

Solution   In the expression $-x^2$ it is the quantity x which is squared, so that $-x^2 = -(x \times x)$. On the other hand $(-x)^2$ means $(-x) \times (-x)$ which equals $+x^2$. The two expressions are not the same.

Recall that when a negative number is multiplied by a negative number the result is positive

Following the previous two examples we emphasize again the importance of the position of brackets in an expression.

Self-assessment questions 4.2

1. Explain the meaning of the terms 'power' and 'base'.
2. What is meant by an index?
3. Explain the distinction between $(xyz)^2$ and xyz^2.
4. Explain the distinction between $(-3)^4$ and -3^4.

Exercise 4.2

1. Evaluate the following without using a calculator: 2^4, $\left(\frac{1}{2}\right)^2$, 1^8, 3^5 and 0^3.

2. Evaluate 10^4, 10^5 and 10^6 without using a calculator.

3. Use a calculator to evaluate 11^4, 16^8, 39^4 and 1.5^7.

4. Write out fully (a) $a^4 b^2 c$ and (b) $xy^2 z^4$.

5. Write the following expressions compactly using indices:

   (a) $xxxyyx$ (b) $xxyyzzz$
   (c) $xyzxyz$ (d) $abccba$

6. Using a calculator, evaluate

   (a) 7^4 (b) 7^5 (c) $7^4 \times 7^5$ (d) 7^9
   (e) 8^3 (f) 8^7 (g) $8^3 \times 8^7$ (h) 8^{10}

   Can you spot a rule for multiplying numbers with powers?

7. Without using a calculator, find $(-3)^3$, $(-2)^2$, $(-1)^7$ and $(-1)^4$.

8. Use a calculator to find $(-16.5)^3$, $(-18)^2$ and $(-0.5)^5$.

4.3   Substitution and formulae

Substitution means replacing letters by actual numerical values.

Worked examples

4.11   Find the value of a^4 when $a = 3$.

Solution a^4 means $a \times a \times a \times a$. When we **substitute** the number 3 in place of the letter a we find 3^4 or $3 \times 3 \times 3 \times 3$, that is, 81.

4.12   Find the value of $a + 7b + 3c$ when $a = 1$, $b = 2$ and $c = 3$.

Solution   Letting $b = 2$ we note that $7b = 14$. Letting $c = 3$ we note that $3c = 9$. Therefore, with $a = 1$,

$$a + 7b + 3c = 1 + 14 + 9 = 24$$

4.13   If $x = 4$, find the value of (a) $8x^3$ and (b) $(8x)^3$.

Solution   (a) Substituting $x = 4$ into $8x^3$ we find $8 \times 4^3 = 8 \times 64 = 512$.
(b) Substituting $x = 4$ into $(8x)^3$ we obtain $(32)^3 = 32768$. Note that the use of brackets makes a significant difference to the result.

4.14   Evaluate mk, mn and nk when $m = 5$, $n = -4$ and $k = 3$.

Solution $mk = 5 \times 3 = 15$. Similarly $mn = 5 \times (-4) = -20$ and $nk = (-4) \times 3 = -12$.

4.15   Find the value of $-7x$ when (a) $x = 2$ and (b) $x = -2$.

Solution   (a) Substituting $x = 2$ into $-7x$ we find -7×2 which equals -14.
(b) Substituting $x = -2$ into $-7x$ we find -7×-2 which equals 14.

A **formula** is used to relate two or more quantities. You may already be familiar with the common formula used to find the area of a rectangle:

$$\text{area} = \text{length} \times \text{breadth}$$

In symbols, writing A for area, l for length and b for breadth we have

$$A = l \times b \qquad \text{or simply } A = lb$$

If we are now given particular numerical values for l and b we can use this formula to find A.

Worked examples

4.16   Use the formula $A = lb$ to find A when $l = 10$ and $b = 2.5$.

Solution   Substituting the values $l = 10$ and $b = 2.5$ into the formula $A = lb$ we find $A = 10 \times 2.5 = 25$.

4.17   The formula $V = IR$ is used by electrical engineers. Find the value of V when $I = 12$ and $R = 7$.

Solution   Substituting $I = 12$ and $R = 7$ in $V = IR$ we find $V = 12 \times 7 = 84$.

4.18   Use the formula $y = x^2 + 3x + 4$ to find y when $x = -2$.

Solution   Substituting $x = -2$ into the formula gives

$$y = (-2)^2 + 3(-2) + 4 = 4 - 6 + 4 = 2$$

Self-assessment questions 4.3

1. What is the distinction between an algebraic expression and a formula?

Exercise 4.3

1. Evaluate $3x^2y$ when $x = 2$ and $y = 5$.

2. Evaluate $8x + 17y - 2z$ when $x = 6$, $y = 1$ and $z = -2$.

3. The area A of a circle is found from the formula $A = \pi r^2$ where r is the length of the radius. Taking π to be 3.142 find the areas of the circles whose radii, in centimetres, are (a) $r = 10$, (b) $r = 3$, (c) $r = 0.2$.

4. Evaluate $3x^2$ and $(3x)^2$ when $x = 4$.

5. Evaluate $5x^2$ and $(5x)^2$ when $x = -2$.

6. If $y = 4.85$ find

    (a) $7y$ (b) y^2 (c) $5y + 2.5$ (d) $y^3 - y$

7. If $a = 12.8$, $b = 3.6$ and $c = 9.1$ find

    (a) $a + b + c$ (b) ab (c) bc (d) abc

8. If $C = \frac{5}{9}(F - 32)$, find C when $F = 100$.

9. Evaluate (a) x^2, (b) $-x^2$ and (c) $(-x)^2$, when $x = 7$.

10. Evaluate $x^2 - 7x + 2$ when $x = -9$.

11. Evaluate $2x^2 + 3x - 11$ when $x = -3$.

12. Evaluate $-x^2 + 3x - 5$ when $x = -1$.

13. Evaluate $-9x^2 + 2x$ when $x = 0$.

14. Evaluate $5x^2 + x + 1$ when (a) $x = 3$, (b) $x = -3$, (c) $x = 0$, (d) $x = -1$.

Test and assignment exercises 4

1. Using a calculator, evaluate 44^3, 0.44^2 and 32.5^3.

2. Write the following compactly using indices:

    (a) $xxxyyyyy$ (b) $\dfrac{xxx}{yyyy}$ (c) a^2baab

3. Evaluate the expression $4x^3yz^2$ when $x = 2$, $y = 5$ and $z = 3$.

4. The circumference C of a circle which has a radius of length r is given by the formula $C = 2\pi r$. Find the circumference of the circle with radius 0.5 cm. Take $\pi = 3.142$.

5. Find (a) $21^2 - 16^2$, (b) $(21 - 16)^2$. Comment upon the result.

6. If $x = 4$ and $y = -3$, evaluate

    (a) xy (b) $\dfrac{x}{y}$ (c) $\dfrac{x^2}{y^2}$ (d) $\left(\dfrac{x}{y}\right)^2$

7. Evaluate $2x(x + 4)$ when $x = 7$.

8. Evaluate $4x^2 + 7x$ when $x = 9$.

9. Evaluate $3x^2 - 7x + 12$ when $x = -2$.

10. Evaluate $-x^2 - 11x + 1$ when $x = -3$.

11. The formula $I = V/R$ is used by engineers. Find I when $V = 10$ and $R = 0.01$.

12. Given the formula $A = 1/x$, find A when (a) $x = 1$, (b) $x = 2$, (c) $x = 3$.

13. From the formula $y = 1/(x^2 + x)$ find y when (a) $x = 1$, (b) $x = -1$, (c) $x = 3$.

14. Find the value of $(-1)^n$ (a) when n is an even natural number and (b) when n is an odd natural number.

15. Find the value of $(-1)^{n+1}$ (a) when n is an even natural number and (b) when n is an odd natural number.

5 Indices

Objectives

This chapter

- states the three laws of manipulating indices

- shows how expressions involving indices can be simplified using the three laws

- explains square roots, cube roots and fractional powers

- explains 'scientific notation' for representing very large and very small numbers concisely

5.1 The laws of indices

Recall from Chapter 4 that an index is simply a power and that the plural of index is indices. Expressions involving indices can often be simplified if use is made of the **laws of indices**.

The first law

$$a^m \times a^n = a^{m+n}$$

In words, this states that if two numbers involving the same base but possibly different indices are to be multiplied together, their indices are added. Note that this law can only be applied if both bases are the same.

KEY POINT

The first law: $a^m \times a^n = a^{m+n}$.

Worked examples

5.1 Use the first law of indices to simplify $a^4 \times a^3$.

Solution Using the first law we have $a^4 \times a^3 = a^{4+3} = a^7$. Note that the same result could be obtained by actually writing out all the terms:
$a^4 \times a^3 = (a \times a \times a \times a) \times (a \times a \times a) = a^7$.

5.2   Use the first law of indices to simplify $3^4 \times 3^5$.

Solution   From the first law $3^4 \times 3^5 = 3^{4+5} = 3^9$.

5.3   Simplify $a^4 a^7 b^2 b^4$.

Solution $a^4 a^7 b^2 b^4 = a^{4+7} b^{2+4} = a^{11} b^6$. Note that only those quantities with the same base can be combined using the first law.

5.4   Simplify $4x^2 \times 7x^5$.

Solution   Recall that when multiplying, the order in which we write down the terms does not matter. Therefore we can write

$$4x^2 \times 7x^5 = 4 \times 7 \times x^2 \times x^5$$

which equals $28x^{2+5}$, that is, $28x^7$.

The second law

$$\frac{a^m}{a^n} = a^{m-n}$$

In words, this states that if two numbers involving the same base but possibly different indices are to be divided, their indices are subtracted.

KEY POINT

> The second law: $\dfrac{a^m}{a^n} = a^{m-n}$.

Worked examples

5.5   Use the second law of indices to simplify a^5/a^3.

Solution   The second law states that we subtract the indices, that is, $a^5/a^3 = a^{5-3} = a^2$.

5.6   Use the second law of indices to simplify $3^7/3^4$.

Solution   From the second law $3^7/3^4 = 3^{7-4} = 3^3$.

5.7   Using the second law of indices, simplify x^3/x^3.

Solution   Using the second law of indices we have

$$x^3/x^3 = x^{3-3} = x^0$$

However, note that any expression divided by itself equals 1, and so x^3/x^3 must equal 1. We can conclude from this that any number raised to the power 0 equals 1.

KEY POINT

Any number raised to the power 0 equals 1, that is, $a^0 = 1$.

Worked examples

5.8   Evaluate (a) 14^0 (b) 0.5^0.

Solution   (a) Any number to the power zero equals 1 and so $14^0 = 1$.

(b) Similarly, $0.5^0 = 1$.

5.9   Simplify (a) $\dfrac{12x^4}{3x^2}$ (b) $\dfrac{16x^5y^5}{12x^3y^2}$ (c) $\dfrac{4x^3y^2z}{2x^2y}$

Solution   (a) We note that $12/3 = 4$. Using the second law of indices we have $x^4/x^2 = x^{4-2} = x^2$. Hence

$$\frac{12x^4}{3x^2} = 4x^2$$

(b) We note that $16/12 = 4/3$. Also $x^5/x^3 = x^{5-3} = x^2$ and $y^5/y^2 = y^{5-2} = y^3$. Hence

$$\frac{16x^5y^5}{12x^3y^2} = \frac{4x^2y^3}{3}$$

(c) Note that $4/2 = 2$ and also that $x^3/x^2 = x^{3-2} = x^1 = x$. Further, since y can be written as y^1 we have

$$\frac{y^2}{y} = \frac{y^2}{y^1} = y^{2-1} = y^1 = y$$

Then

$$\frac{4x^3y^2z}{2x^2y} = 2xyz$$

The third law

$$(a^m)^n = a^{mn}$$

If a number is raised to a power, and the result is itself raised to a power, then the two powers are multiplied together.

KEY POINT

The third law: $(a^m)^n = a^{mn}$.

Worked examples

5.10   Simplify $(3^2)^4$.

Solution   The third law states that the two powers are multiplied:
$(3^2)^4 = 3^{2 \times 4} = 3^8$.

5.11   Simplify $(x^4)^3$.

Solution   Using the third law $(x^4)^3 = x^{4 \times 3} = x^{12}$.

5.12   Remove the brackets from the expression $(2a^2)^3$.

Solution $(2a^2)^3$ means $(2a^2) \times (2a^2) \times (2a^2)$. We can write this as

$$2 \times 2 \times 2 \times a^2 \times a^2 \times a^2$$

or simply $8a^6$. We could obtain the same result by noting that both terms in the bracket, that is, the 2 and the a^2, must be raised to the power 3, that is,

$$(2a^2)^3 = 2^3(a^2)^3 = 8a^6$$

The result of the previous example can be generalized to any term of the form $(a^m b^n)^k$. To simplify such an expression we make use of the formula $(a^m b^n)^k = a^{mk} b^{nk}$

KEY POINT

$$(a^m b^n)^k = a^{mk} b^{nk}$$

Worked example

5.13   Remove the brackets from the expression $(x^2y^3)^4$.

Solution   Using the previous result we find

$$(x^2y^3)^4 = x^8y^{12}$$

Self-assessment questions 5.1

1. State the three laws of indices.
2. Explain what is meant by a^0.
3. Explain what is meant by x^1.

Exercise 5.1

1. Simplify
   (a) $5^7 \times 5^{13}$ (b) $9^8 \times 9^5$
   (c) $11^2 \times 11^3 \times 11^4$

2. Simplify
   (a) $\dfrac{15^3}{15^2}$ (b) $\dfrac{4^{18}}{4^9}$ (c) $\dfrac{5^{20}}{5^{19}}$

3. Simplify
   (a) a^7a^3 (b) a^4a^5 (c) $b^{11}b^{10}b$

4. Simplify
   (a) $x^7 \times x^8$ (b) $y^4 \times y^8 \times y^9$
   (c) $3x^6 \times 5x^9$

5. Explain why the laws of indices cannot be used to simplify $19^8 \times 17^8$.

6. Simplify
   (a) $(7^3)^2$ (b) $(4^2)^8$ (c) $(7^9)^2$

7. Simplify $\dfrac{1}{(5^3)^8}$

8. Simplify
   (a) $(x^2y^3)(x^3y^2)$ (b) $(2x^2)(3x^4)$

   (c) $(a^2bc^2)(b^2ca)$ (d) $\dfrac{6c^3d^3}{3cd^2}$

5.2   Negative powers

Sometimes a number is raised to a negative power. This is interpreted as follows:

$$a^{-m} = \frac{1}{a^m}$$

This can also be rearranged and expressed in the form

$$a^m = \frac{1}{a^{-m}}$$

KEY POINT

$$a^{-m} = \frac{1}{a^m}, \qquad a^m = \frac{1}{a^{-m}}$$

For example,

$$3^{-2} \text{ means } \frac{1}{3^2}, \text{ that is, } \frac{1}{9}$$

Similarly,

$$\text{the number } \frac{1}{5^{-2}} \text{ can be written } 5^2, \text{ or simply } 25$$

To see the justification for this, note that because any number raised to the power 0 equals 1 we can write

$$\frac{1}{a^m} = \frac{a^0}{a^m}$$

Using the second law of indices to simplify the right-hand side we obtain $a^0/a^m = a^{0-m} = a^{-m}$ so that $1/a^m$ is the same as a^{-m}.

Worked examples

5.14   Evaluate

(a) 2^{-5} (b) $\dfrac{1}{3^{-4}}$

Solution   (a) $2^{-5} = \dfrac{1}{2^5} = \dfrac{1}{32}$

(b) $\dfrac{1}{3^{-4}} = 3^4$ or simply 81

5.15   Evaluate (a) 10^{-1} (b) 10^{-2}

Solution   (a) 10^{-1} means $1/10^1$, or simply $1/10$. It is important to recognize that 10^{-1} is therefore the same as 0.1.

(b) 10^{-2} means $1/10^2$ or $1/100$. So 10^{-2} means the same as 0.01.

Self-assessment questions 5.2

1. Explain how the negative power in a^{-m} is interpreted.

Exercise 5.2

1. Simplify

   (a) $a^{13}a^{-2}$ (b) $x^{-9}x^{-7}$ (c) $x^{-21}x^2x$
   (d) $(4^{-3})^2$

2. Evaluate

   (a) 10^{-3} (b) 10^{-4} (c) 10^{-5}

3. Evaluate $4^{-8}/4^{-6}$ and $3^{-5}/3^{-8}$ without using a calculator.

5.3   Square roots, cube roots and fractional powers

Consider the relationship between the numbers 5 and 25. We know that $5^2 = 25$ so that 25 is the square of 5. Equivalently we can say that 5 is a **square root** of 25. The symbol $\sqrt{\ }$ is used to denote a square root and we write

$$5 = \sqrt{25}$$

We also note that $(-5)(-5) = (-5)^2 = 25$ and so -5 is a square root of 25. Hence we can write

$$-5 = \sqrt{25}$$

We can write both results together by using the 'plus or minus' sign $\pm$. We write

$$\sqrt{25} = \pm 5$$

In general, a **square root** of a number is a number which when squared gives the given number.

Similarly, since $8^2 = 64$ and $(-8)^2 = 64$ we can write $\sqrt{64} = \pm 8$. Note that there are two square roots of any positive number, but negative numbers possess no square roots.

The **cube root** of a number is the number which when cubed gives the given number. The symbol for a cube root is $\sqrt[3]{\ }$. So, for example, since $2^3 = 8$ we can write $\sqrt[3]{8} = 2$. As another example, we note $(-2)^3 = -8$

and hence $\sqrt[3]{-8} = -2$. All numbers, both positive and negative, possess a single cube root.

Other roots can be defined in a similar way. For example, because $3^4 = 81$ we can write $\sqrt[4]{81} = 3$. Also since $(-3)^4 = 81$, then $\sqrt[4]{81} = -3$.

Most calculators enable you to calculate square roots although only the positive value is normally given. Many will also calculate cube and other roots. You should ensure that you can use your calculator correctly to determine square and cube roots.

Sometimes a number is raised to a fractional power. This is interpreted as follows:

KEY POINT

$$a^{1/n} = \sqrt[n]{a}$$

For example $a^{1/3} = \sqrt[3]{a}$, that is, the cube root of a number can be expressed as that number to the power 1/3. Similarly $a^{1/2} = \sqrt{a}$.

Note that using the third law we can write $(a^{1/2})^2 = a^{(1/2)\times 2} = a^1 = a$, so that squaring the square root results in a. Similarly, cubing the cube root of a number results in the original number being returned, that is, $(a^{1/3})^3 = a$.

Worked examples

5.16   Evaluate $(81)^{1/2}$.

Solution $(81)^{1/2} = \sqrt{81} = \pm 9$.

5.17   Explain what is meant by the number $27^{1/3}$.

Solution $27^{1/3}$ can be written $\sqrt[3]{27}$, that is, the cube root of 27. The cube root of 27 is 3, since $3 \times 3 \times 3 = 27$, and so $27^{1/3} = 3$. Note also that since $27 = 3^3$ we can write

$$(27)^{1/3} = (3^3)^{1/3} = 3^{(3 \times 1/3)} \quad \text{using the third law}$$
$$= 3^1 \quad = 3$$

The following worked example shows how we deal with negative fractional powers.

5.18   Explain what is meant by the number $(81)^{-1/2}$.

Solution   Recall from our work on negative powers that $a^{-m} = 1/a^m$. Therefore we can write $(81)^{-1/2}$ as $1/(81)^{1/2}$. Now $81^{1/2} = \sqrt{81} = \pm 9$ and so

$$(81)^{-1/2} = \frac{1}{\pm 9} = \pm \frac{1}{9}$$

Self-assessment questions 5.3

1. Explain the meaning of the fractional powers $x^{1/2}$ and $x^{1/3}$.
2. What are the square roots of 100? Explain why the number -100 does not have any square roots.

Exercise 5.3

1. Evaluate

   (a) $64^{1/3}$ (b) $144^{1/2}$ (c) $16^{-1/4}$ (d) $25^{-1/2}$

   (e) $\dfrac{1}{32^{-1/5}}$

2. Simplify and then evaluate

   (a) $(3^{-1/2})^4$ (b) $(8^{1/3})^{-1}$

5.4   Scientific notation

It is often necessary to use very large numbers such as 65000000000 or very small numbers such as 0.000000001. **Scientific notation** can be used to express such numbers in a more concise form which avoids writing very lengthy strings of numbers. Each number is written in the form

$$a \times 10^n$$

where a is usually a number between 1 and 10. We also make use of the fact that

$$10 = 10^1, \quad 100 = 10^2, \quad 1000 = 10^3 \text{ and so on}$$

and also that

$$10^{-1} = \frac{1}{10} = 0.1, \quad 10^{-2} = \frac{1}{100} = 0.01 \text{ and so on}$$

Then, for example,

the number 4000 can be written $4 \times 1000 = 4 \times 10^3$

Similarly

the number 68,000 can be written $6.8 \times 10,000 = 6.8 \times 10^4$

and

the number 0.09 can be written $9 \times 0.01 = 9 \times 10^{-2}$

Note that all three numbers have been written in the form $a \times 10^n$ where a lies between 1 and 10.

Worked examples

5.19   Express the following numbers in scientific notation:

(a) 54 (b) −276 (c) 0.3

Solution   (a) 54 can be written as 5.4×10, so in scientific notation we have 5.4×10^1.

(b) Negative numbers cause no problem: $-276 = -2.76 \times 10^2$.

(c) We can write 0.3 as 3×0.1 or 3×10^{-1}.

5.20   Write out fully the following numbers:

(a) 2.7×10^{-1} (b) 9.6×10^5 (c) -8.2×10^2

Solution   (a) $2.7 \times 10^{-1} = 0.27$.   (b) $9.6 \times 10^5 = 9.6 \times 100000 = 960000$.

(c) $-8.2 \times 10^2 = -8.2 \times 100 = -820$.

5.21   Simplify the expression $(3 \times 10^2) \times (5 \times 10^3)$.

Solution   The order in which the numbers are written down does not matter and so we can write

$$(3 \times 10^2) \times (5 \times 10^3) = 3 \times 5 \times 10^2 \times 10^3 = 15 \times 10^5$$

In the standard form of scientific notation we would write this as 1.5×10^6.

Self-assessment questions 5.4

1. What is the purpose of using scientific notation?

Exercise 5.4

1. Express each of the following numbers in scientific notation:

(a) 45 (b) 45000 (c) −450
(d) 90000000 (e) 0.15 (f) 0.00036
(g) 3.5 (h) −13.2 (i) 1000000
(j) 0.0975 (k) 45.34

2. Write out fully the following numbers:

(a) 3.75×10^2 (b) 3.97×10^1
(c) 1.875×10^{-1} (d) -8.75×10^{-3}

Test and assignment exercises 5

1. Simplify

(a) $\dfrac{z^5}{z^{-5}}$ (b) z^0 (c) $\dfrac{z^8 z^6}{z^{14}}$

2. Evaluate

(a) $0.25^{1/2}$ (b) $(4096)^{1/3}$ (c) $(2601)^{1/2}$
(d) $16^{-1/2}$

3. Simplify $\dfrac{x^8 x^{-3}}{x^{-5} x^2}$

4. Find the value of $(1/7)^0$.

5. Simplify

(a) $(abc^2)^2$ (b) $(xy^2 z^3)^2$ (c) $(8x^2)^{-3}$

6. Express each of the following numbers in scientific notation:

(a) 5792 (b) 98.4 (c) 0.001 (d) −66.667

6 Simplifying algebraic expressions

Objectives	This chapter
	• describes a number of ways in which complicated algebraic expressions can be simplified

6.1 Addition and subtraction of like terms

Like terms are multiples of the same quantity. For example $3y$, $72y$ and $0.5y$ are all multiples of y and so are like terms. Similarly, $5x^2$, $-3x^2$ and $\frac{1}{2}x^2$ are all multiples of x^2 and so are like terms. xy, $17xy$ and $-91xy$ are all multiples of xy and are therefore like terms. Like terms can be collected together and added or subtracted in order to simplify them.

Worked examples

6.1 Simplify $3x + 7x - 2x$.

Solution All three terms are multiples of x and so are like terms. Therefore $3x + 7x - 2x = 8x$.

6.2 Simplify $3x + 2y$.

Solution $3x$ and $2y$ are not like terms. One is a multiple of x and the other is a multiple of y. The expression $3x + 2y$ cannot be simplified.

6.3 Simplify $x + 7x + x^2$.

Solution The like terms are x and $7x$. These can be simplified to $8x$. Then $x + 7x + x^2 = 8x + x^2$. Note that $8x$ and x^2 are not like terms and so this expression cannot be simplified further.

6.4   Simplify $ab + a^2 - 7b^2 + 9ab + 8b^2$.

Solution   The terms ab and $9ab$ are like terms. Similarly the terms $-7b^2$ and $8b^2$ are like terms. These can be collected together and then added or subtracted as appropriate. Thus

$$ab + a^2 - 7b^2 + 9ab + 8b^2 = ab + 9ab + a^2 - 7b^2 + 8b^2 = 10ab + a^2 + b^2$$

Exercise 6.1

1. Simplify, if possible,

   (a) $5p - 10p + 11q + 8q$
   (b) $-7r - 13s + 2r + z$
   (c) $181z + 13r - 2$
   (d) $x^2 + 3y^2 - 2y + 7x^2$
   (e) $4x^2 - 3x + 2x + 9$

2. Simplify

   (a) $5y + 8p - 17y + 9q$
   (b) $7x^2 - 11x^3 + 14x^2 + y^3$

6.2   Multiplication and division

Rules for multiplying and dividing algebraic quantities are the same as those already given for manipulating numbers.

KEY POINT

Recall that when multiplying

$$\text{positive} \times \text{positive} = \text{positive}$$

$$\text{positive} \times \text{negative} = \text{negative}$$

$$\text{negative} \times \text{positive} = \text{negative}$$

$$\text{negative} \times \text{negative} = \text{positive}$$

When dividing

$$\frac{\text{positive}}{\text{positive}} = \text{positive}$$

$$\frac{\text{positive}}{\text{negative}} = \text{negative}$$

$$\frac{\text{negative}}{\text{positive}} = \text{negative}$$

$$\frac{\text{negative}}{\text{negative}} = \text{positive}$$

For example,

$$(a) \times (-b) = -ab$$

$$5a \times 3b = 5 \times 3 \times a \times b = 15ab$$

$$\frac{x}{-y} = -\frac{x}{y} \qquad \frac{-x}{y} = -\frac{x}{y} \qquad \frac{-a}{-b} = \frac{a}{b}$$

Worked examples

6.5   Simplify (a) $(6x)(-2x)$ (b) $(-3y^2)(-2y)$.

Solution   (a) $(6x)(-2x)$ means $(6x) \times (-2x)$ which equals $-12x^2$.
(b) $(-3y^2)(-2y) = (-3y^2) \times (-2y) = 6y^3$.

6.6   Simplify (a) $\dfrac{x^3}{-x^4}$ (b) $\dfrac{-x^{-7}}{-x^5}$

Solution   (a) Dividing a positive quantity by a negative quantity results in another negative quantity. Therefore

$$\frac{x^3}{-x^4} = -\frac{x^3}{x^4}$$

Using the second law of indices we can write this as $-x^{-1}$.

(b) Dividing a negative quantity by a negative quantity results in a positive quantity. Therefore,

$$\frac{-x^{-7}}{-x^5} = \frac{x^{-7}}{x^5}$$

Using the second law of indices we can write this as x^{-12}.

Exercise 6.2

1. Simplify

   (a) $(3z)(-7z)$ (b) $3z - 7z$

2. Simplify

   (a) $(-x)(3x)$ (b) $-x + 3x$

3. Simplify

   (a) $(-2x)(-x)$ (b) $-2x - x$

4. Simplify

   (a) $\dfrac{x^5}{-x^3}$ (b) $\dfrac{-y^3}{y^7}$ (c) $\dfrac{-y}{-y^2}$ (d) $\dfrac{-y^{-3}}{-y^4}$

6.3   Removing brackets

Recall from your study of arithmetic that the expression $(5 - 4) + 7$ is different from $5 - (4 + 7)$ because of the position of the brackets. In order to simplify an expression it is often necessary to remove brackets.

Removing brackets from expressions of the form $a(b + c)$ and $a(b - c)$

In an expression such as $a(b + c)$, it is intended that the a multiplies all the bracketed terms:

| KEY POINT | $a(b+c) = ab + ac$ | Similarly: | $a(b-c) = ab - ac$ |

Worked examples

6.7 Remove the brackets from

(a) $6(x+5)$ (b) $8(2x-4)$

Solution (a) In the expression $6(x+5)$ it is intended that the 6 multiplies both terms in the brackets. Therefore

$$6(x+5) = 6x + 30$$

(b) In the expression $8(2x-4)$ the 8 multiplies both terms in the brackets so that

$$8(2x-4) = 16x - 32$$

6.8 Remove the brackets from the expression $7(5x + 3y)$.

Solution The 7 multiplies both the terms in the bracket. Therefore

$$7(5x + 3y) = 7(5x) + 7(3y) = 35x + 21y$$

6.9 Remove the brackets from $-(x+y)$.

Solution The expression $-(x+y)$ actually means $-1(x+y)$. It is intended that the -1 multiplies both terms in the brackets, therefore

$$-(x+y) = -1(x+y) = (-1) \times x + (-1) \times y = -x - y$$

6.10 Remove brackets from the expression

$$(x+y)z$$

Solution Note that the order in which we write down the terms to be multiplied does not matter, so that we can write $(x+y)z$ as $z(x+y)$. Then

$$z(x+y) = zx + zy$$

Alternatively note that $(x+y)z = xz + yz$ which is an equivalent form of the answer.

6.11 Remove the brackets from the expressions

(a) $5(x-2y)$ (b) $(x+3)(-1)$

Solution (a) $5(x-2y) = 5x - 5(2y) = 5x - 10y$.

(b) $(x+3)(-1) = (-1)(x+3) = -1x - 3 = -x - 3$.

6.12   Simplify $x + 8(x - y)$.

Solution   An expression such as this is simplified by first removing the brackets and then collecting together like terms. Removing the brackets we find

$$x + 8(x - y) = x + 8x - 8y$$

Collecting like terms we obtain $9x - 8y$.

6.13   Remove the brackets from

$$\text{(a) } \tfrac{1}{2}(x + 2) \text{ (b) } \tfrac{1}{2}(x - 2) \text{ (c) } -\tfrac{1}{3}(a + b)$$

Solution   (a) In the expression $\tfrac{1}{2}(x + 2)$ it is intended that the $\tfrac{1}{2}$ multiplies both the terms in the brackets. So

$$\frac{1}{2}(x + 2) = \frac{1}{2}x + \frac{1}{2}(2) = \frac{1}{2}x + 1$$

(b) Similarly,

$$\frac{1}{2}(x - 2) = \frac{1}{2}x - \frac{1}{2}(2) = \frac{1}{2}x - 1$$

(c) In the expression $-\tfrac{1}{3}(a + b)$, the term $-\tfrac{1}{3}$ multiplies both terms in the brackets. So

$$-\frac{1}{3}(a + b) = -\frac{1}{3}a - \frac{1}{3}b$$

Removing brackets from expressions of the form $(a + b)(c + d)$

Here it is intended that the quantity $(a + b)$ multiplies both the c and the d. Therefore

$$(a + b)(c + d) = (a + b)c \quad + \quad (a + b)d$$

Each of these two terms can be expanded further to give

$$(a + b)c = ac + bc \quad \text{and} \quad (a + b)d = ad + bd$$

Therefore:

KEY POINT $(a + b)(c + d) = ac + bc + ad + bd$

Worked examples

6.14 Remove the brackets from $(3 + x)(2 + y)$.

Solution

$$(3 + x)(2 + y) = (3 + x)(2) + (3 + x)y$$
$$= 6 + 2x + 3y + xy$$

6.15 Remove the brackets from $(x + 6)(x - 3)$.

Solution

$$(x + 6)(x - 3) = (x + 6)x + (x + 6)(-3)$$
$$= x^2 + 6x - 3x - 18$$
$$= x^2 + 3x - 18$$

6.16 Remove the brackets from

(a) $(1 - x)(2 - x)$ (b) $(-x - 2)(2x - 1)$

Solution (a) $(1 - x)(2 - x) = (1 - x)2 + (1 - x)(-x)$
$$= 2 - 2x - x + x^2$$
$$= 2 - 3x + x^2$$

(b) $(-x - 2)(2x - 1) = (-x - 2)(2x) + (-x - 2)(-1)$
$$= -2x^2 - 4x + x + 2$$
$$= -2x^2 - 3x + 2$$

6.17 Remove the brackets from the expression $3(x + 1)(x - 1)$.

Solution First consider the expression $(x + 1)(x - 1)$.
$$(x + 1)(x - 1) = (x + 1)x + (x + 1)(-1)$$
$$= x^2 + x - x - 1$$
$$= x^2 - 1$$

Then $3(x + 1)(x - 1) = 3(x^2 - 1) = 3x^2 - 3$.

Exercise 6.3

1. Remove the brackets from

   (a) $4(x+1)$ (b) $-4(x+1)$ (c) $4(x-1)$
   (d) $-4(x-1)$

2. Remove the brackets from the following expressions:

   (a) $5(x-y)$ (b) $19(x+3y)$
   (c) $8(a+b)$ (d) $(5+x)y$ (e) $12(x+4)$
   (f) $17(x-9)$ (g) $-(a-2b)$
   (h) $\frac{1}{2}(2x+1)$ (i) $-3m(-2+4m+3n)$

3. Remove the brackets and simplify the following:

   (a) $18-13(x+2)$ (b) $x(x+y)$

4. Remove the brackets and simplify the following expressions:

   (a) $(x+1)(x+6)$ (b) $(x+4)(x+5)$
   (c) $(x-2)(x+3)$ (d) $(x+6)(x-1)$
   (e) $(x+y)(m+n)$ (f) $(4+y)(3+x)$
   (g) $(5-x)(5+x)$ (h) $(17x+2)(3x-5)$

5. Remove the brackets and simplify the following expressions:

   (a) $(x+3)(x-7)$ (b) $(2x-1)(3x+7)$
   (c) $(4x+1)(4x-1)$ (d) $(x+3)(x-3)$
   (e) $(2-x)(3+2x)$

6. Remove the brackets and simplify the following expressions:

   (a) $\frac{1}{2}(x+2y)+\frac{7}{2}(4x-y)$

   (b) $\frac{3}{4}(x-1)+\frac{1}{4}(2x+8)$

7. Remove the brackets from

   (a) $-(x-y)$ (b) $-(a+2b)$

   (c) $-\frac{1}{2}(3p+q)$

8. Remove the brackets from
   $(x+1)(x+2)$. Use your result to remove the brackets from
   $(x+1)(x+2)(x+3)$.

Test and assignment exercises 6

1. Simplify

   (a) $7x^2+4x^2+9x-8x$
   (b) $y+7-18y+1$
   (c) $a^2+b^2+a^3-3b^2$

2. Simplify

   (a) $(3a^2b)\times(-a^3b^2c)$ (b) $\dfrac{x^3}{-x^2}$

3. Remove the brackets from

   (a) $(a+3b)(7a-2b)$ (b) $x^2(x+2y)$
   (c) $x(x+y)(x-y)$

4. Remove the brackets from

   (a) $(7x+2)(3x-1)$ (b) $(1-x)(x+3)$
   (c) $(5+x)x$ (d) $(8x+4)(7x-2)$

5. Remove the brackets and simplify

   (a) $3x(x+2)-7x^2$
   (b) $-(2a+3b)(a+b)$
   (c) $4(x+7)+13(x-2)$
   (d) $5(2a+5)-3(5a-2)$
   (e) $\frac{1}{2}(a+4b)+\frac{3}{2}a$

7 Factorization

Objectives

This chapter

- explains what is meant by the 'factors' of an algebraic expression
- shows how an algebraic expression can be factorized

7.1 Factors and common factors

Recall from Chapter 1 that a number is **factorized** when it is written as a product. For example, 15 may be factorized into 3×5. We say that 3 and 5 are **factors** of 15. The number 16 can be written as 8×2 or 4×4, or even as 16×1, and so the factorization may not be unique.

Algebraic expressions can also be factorized. Consider the expression $5x + 15y$. Both $5x$ and $15y$ have the number 5 common to both terms. We say that 5 is a **common factor**. Any common factors can be written outside a bracketed term. Thus $5x + 15y = 5(x + 3y)$. Removal of the brackets will result in the original expression and can always be used to check your answer. We see that factorization can be thought of as reversing the process of removing brackets. Similarly, if we consider the expression $x^2 + 2x$, we note that both terms contain the factor x, and so $x^2 + 2x$ can be written as $x(x + 2)$. Hence x and $x + 2$ are both factors of $x^2 + 2x$.

Worked examples

7.1 Factorize $3x + 12$.

Solution The number 12 can be factorized as 3×4 so that 3 is a common factor of $3x$ and 12. We can write $3x + 12 = 3x + 3(4)$. Any common factors are written in front of a bracket and the contents of the bracket are adjusted accordingly. So

$$3x + 3(4) = 3(x + 4)$$

Note again that this answer can be checked by removing the brackets.

7.2   List the ways in which $15x^2$ can be written as a product of its factors.

Solution $15x^2$ can be written in many different ways. These are $15x^2 \times 1$, $15x \times x$, $15 \times x^2$, $5x \times 3x$, $5 \times 3x^2$, and $3 \times 5x^2$.

7.3   Factorize $8x^2 - 12x$.

Solution   We can write
$$8x^2 - 12x = (4x)(2x) - (4x)3$$
so that both terms contain the factor $4x$. This is placed at the front of a bracket to give
$$8x^2 - 12x = 4x(2x - 3)$$

7.4   What factors are common to the terms $5x^2$ and $15x^3$? Factorize $5x^2 + 15x^3$.

Solution   Both terms contain a factor of 5. Because x^3 can be written $x^2 \times x$, both $5x^2$ and $15x^3$ contain a factor x^2. Therefore
$$5x^2 + 15x^3 = 5x^2 + (5x^2)(3x) = 5x^2(1 + 3x)$$

7.5   Factorize $6x + 3x^2 + 9xy$.

Solution   By careful inspection of all of the terms we see that $3x$ is a factor of each term. Hence
$$6x + 3x^2 + 9xy = 3x(2 + x + 3y)$$
Hence the factors of $6x + 3x^2 + 9xy$ are $3x$ and $2 + x + 3y$.

Self-assessment questions 7.1

1. Explain what is meant by 'factorizing an expression'.

Exercise 7.1

1. Remove the brackets from
   (a) $9(x + 3)$ (b) $-5(x - 2)$ (c) $\frac{1}{2}(x + 1)$
   (d) $-(a - 3b)$ (e) $\frac{1}{2(x + y)}$ (f) $\frac{x}{y(x - y)}$

2. List all the factors of each of (a) $4x^2$, (b) $6x^3$.

3. Factorize
   (a) $10x + 20y$ (b) $12a + 3b$ (c) $4x - 6xy$
   (d) $7a + 14$ (e) $10m - 15$
   (f) $\frac{1}{5a + 35b}$ (g) $\frac{1}{5a^2 + 35ab}$

4. Factorize
   (a) $15x^2 + 3x$ (b) $4x^2 - 3x$ (c) $4x^2 - 8x$
   (d) $15 - 3x^2$ (e) $10x^3 + 5x^2 + 15x^2y$
   (f) $6a^2b - 12ab^2$ (g) $16abc - 8ab^2 + 24bc$

7.2   Factorizing quadratic expressions

Expressions of the form $ax^2 + bx + c$ where a, b, and c are numbers are called **quadratic expressions**. The numbers b or c may equal zero but a must not be zero. The number a is called the **coefficient** of x^2, b is the coefficient of x and c is called the **constant term**.
We see that

$$2x^2 + 3x - 1, \quad x^2 + 3x + 2, \quad x^2 + 7 \text{ and } 2x^2 - x$$

are all quadratic expressions.

KEY POINT

> An expression of the form $ax^2 + bx + c$ where a, b and c are numbers is called a quadratic expression. The coefficient of x^2 is a, the coefficient of x is b and the constant term is c.

To factorize such an expression means to express it as a product of two terms. For example, removing the brackets from $(x + 3)(x + 2)$ gives $x^2 + 5x + 6$. Reversing the process, $x^2 + 5x + 6$ can be factorized to $(x + 3)(x + 2)$. Not all quadratic expressions can be factorized in this way. We shall now explore how such factorization is attempted.

Quadratic expressions where the coefficient of x^2 is 1

Consider the expression $(x + m)(x + n)$. Removing the brackets we find

$$
\begin{aligned}
(x + m)(x + n) &= (x + m)x \quad + \quad (x + m)n \\
&= x^2 + mx + nx + mn \\
&= x^2 + (m + n)x + mn
\end{aligned}
$$

Note that the coefficient of the x term is the sum $m + n$ and the constant term is the product mn. Using this information several quadratic expressions can be factorized by careful inspection. For example, suppose we wish to factorize $x^2 + 5x + 6$. We know that $x^2 + (m + n)x + mn$ can be factorized to $(x + m)(x + n)$. We seek values of m and n so that

$$x^2 + 5x + 6 = x^2 + (m + n)x + mn$$

Comparing the coefficients of x on both sides we require

$$5 = m + n$$

Comparing the constant terms on both sides we require

$$6 = mn$$

By inspection we see that $m = 3$ and $n = 2$ have this property and so

$$x^2 + 5x + 6 = (x + 3)(x + 2)$$

Note that the answer can be easily checked by removing the brackets again.

Worked examples

7.6   Factorize the quadratic expression $x^2 + 8x + 12$.

Solution   The factorization of $x^2 + 8x + 12$ will be of the form $(x + m)(x + n)$. This means that mn must equal 12 and $m + n$ must equal 8. The two numbers must therefore be 2 and 6. So

$$x^2 + 8x + 12 = (x + 2)(x + 6)$$

Note again that the answer can be checked by removing the brackets.

7.7   Factorize $x^2 + 10x + 25$.

Solution   We try to factorize in the form $(x + m)(x + n)$. We require $m + n$ to equal 10 and mn to equal 25. If $m = 5$ and $n = 5$ this requirement is met. Therefore $x^2 + 10x + 25 = (x + 5)(x + 5)$. It is usual practice to write this as $(x + 5)^2$.

7.8   Factorize $x^2 - 121$.

Solution   In this example the x term is missing. We still attempt to factorize as $(x + m)(x + n)$. We require $m + n$ to equal 0 and mn to equal -121. Some thought shows that if $m = 11$ and $n = -11$ this requirement is met. Therefore $x^2 - 121 = (x + 11)(x - 11)$.

7.9   Factorize $x^2 - 5x + 6$.

Solution   We try to factorize in the form $(x + m)(x + n)$. We require $m + n$ to equal -5 and mn to equal 6. By inspection we see that if $m = -3$ and $n = -2$ this requirement is met. Therefore $x^2 - 5x + 6 = (x - 3)(x - 2)$.

Quadratic expressions where the coefficient of x^2 is not 1

These expressions are a little harder to factorize. All possible factors of the first and last terms must be found and various combinations of these should be attempted until the required answer is found. This involves trial and error along with educated guesswork and practice.

Worked examples

7.10   Factorize, if possible, the expression $2x^2 + 11x + 12$.

Solution   The factors of the first term, $2x^2$, are $2x$ and x. The factors of the last term, 12, are

$$12, 1 \quad -12, -1 \quad 6, 2 \quad 6, -2 \quad \text{and} \quad 4, 3 \quad -4, -3$$

We can try each of these combinations in turn to find which gives us a coefficient of x of 11. For example, removing the brackets from

$$(2x + 12)(x + 1)$$

gives

$$(2x + 12)(x + 1) = (2x + 12)x + (2x + 12)(1)$$
$$= 2x^2 + 12x + 2x + 12$$
$$= 2x^2 + 14x + 12$$

which has an incorrect middle term. By trying further combinations it turns out that the only one producing a middle term of $11x$ is $(2x + 3)(x + 4)$ because

$$(2x + 3)(x + 4) = (2x + 3)(x) + (2x + 3)(4)$$
$$= 2x^2 + 3x + 8x + 12$$
$$= 2x^2 + 11x + 12$$

so that $(2x + 3)(x + 4)$ is the correct factorization.

7.11   Factorize, if possible, $4x^2 + 6x + 2$.

Solution   Before we try to factorize this quadratic expression notice that there is a factor of 2 in each term so that we can write it as $2(2x^2 + 3x + 1)$. Now consider the quadratic expression $2x^2 + 3x + 1$. The factors of the first term, $2x^2$, are $2x$ and x. The factors of the last term, 1, are simply 1 and 1, or -1 and -1. We can try these combinations in turn to find which gives us a middle term of $3x$. Removing the brackets from $(2x + 1)(x + 1)$ gives $2x^2 + 3x + 1$ which has the correct middle term. Finally, we can write

$$4x^2 + 6x + 2 = 2(2x^2 + 3x + 1) = 2(2x + 1)(x + 1)$$

7.12   Factorize $6x^2 + 7x - 3$.

Solution   The first term may be factorized as $6x \times x$ and also as $3x \times 2x$. The factors of the last term are

$$3, -1 \qquad \text{and} \qquad -3, 1$$

We need to try each combination in turn to find which gives us a coefficient of x of 7. For example, removing the brackets from

$$(6x + 3)(x - 1)$$

gives

$$(6x + 3)(x - 1) = (6x + 3)x + (6x + 3)(-1)$$
$$= 6x^2 + 3x - 6x - 3$$
$$= 6x^2 - 3x - 3$$

which has an incorrect middle term. By trying further combinations it turns out that the only one producing a middle term of $7x$ is $(3x - 1)(2x + 3)$ because

$$(3x - 1)(2x + 3) = (3x - 1)2x + (3x - 1)(3)$$
$$= 6x^2 - 2x + 9x - 3$$
$$= 6x^2 + 7x - 3$$

The correct factorization is therefore $(3x - 1)(2x + 3)$.

Until you have sufficient experience at factorizing quadratic expressions you must be prepared to go through the process of trying all possible combinations until the correct answer is found.

Self-assessment questions 7.2

1. Not all quadratic expressions can be factorized. Try to find an example of one such expression.

Exercise 7.2

1. Factorize the following quadratic expressions:

   (a) $x^2 + 3x + 2$ (b) $x^2 + 13x + 42$
   (c) $x^2 + 2x - 15$ (d) $x^2 + 9x - 10$
   (e) $x^2 - 11x + 24$ (f) $x^2 - 100$
   (g) $x^2 + 4x + 4$

2. Factorize the following quadratic expressions:

   (a) $2x^2 - 5x - 3$ (b) $3x^2 - 5x - 2$
   (c) $10x^2 + 11x + 3$ (d) $2x^2 + 12x + 16$

3. Factorize the following quadratic expressions:

(a) $x^2 + 3x - 10$ (b) $2x^2 - 3x - 20$
(c) $9x^2 - 1$ (d) $10x^2 + 14x - 12$
(e) $x^2 + 15x + 26$ (f) $-x^2 - 2x + 3$

Test and assignment exercises 7

1. Factorize the following expressions:

(a) $7x + 49$ (b) $121x + 22y$ (c) $a^2 + ab$
(d) $ab + b^2$ (e) $ab^2 + ba^2$

2. Factorize the following quadratic expressions:

(a) $3x^2 + x - 2$ (b) $x^2 - 144$
(c) $s^2 - 5s + 6$ (d) $2y^2 - y - 15$

3. Factorize the following:

(a) $1 - x^2$ (b) $x^2 - 1$ (c) $9 - x^2$
(d) $x^2 - 81$ (e) $25 - y^2$

4. Factorize the denominators of the following expressions:

(a) $\dfrac{1}{x^2 + 6x}$ (b) $\dfrac{3}{s^2 + 3s + 2}$

(c) $\dfrac{3}{s^2 + s - 2}$ (d) $\dfrac{5}{x^2 + 11x + 28}$

(e) $\dfrac{x}{2x^2 - 17x - 9}$

8 Algebraic fractions

Objectives	This chapter
	• explains how to simplify algebraic fractions by cancelling common factors
	• explains how algebraic fractions can be multiplied and divided
	• explains how algebraic fractions can be added and subtracted

8.1 Introduction

Just as one whole number divided by another is a numerical fraction, one algebraic expression divided by another is called an **algebraic fraction.**

$$\frac{x}{y} \qquad \frac{x^2 + y}{x} \qquad \frac{3x + 2}{7}$$

are all examples of algebraic fractions. The top line is known as the **numerator** of the fraction, and the bottom line is the **denominator**.

8.2 Cancelling common factors

Consider the numerical fraction $\frac{3}{12}$. To simplify this we factorize both the numerator and the denominator. Any factors which appear in both the numerator and the denominator are called **common factors**. These can be cancelled. For example,

$$\frac{3}{12} = \frac{1 \times 3}{4 \times 3} = \frac{1 \times \cancel{3}}{4 \times \cancel{3}} = \frac{1}{4}$$

The same process is applied when dealing with algebraic fractions.

Worked examples

8.1 Simplify $\dfrac{5}{25 + 15x}$

Solution First of all note that the denominator can be factorized as $5(5 + 3x)$. There is therefore a factor of 5 in both the numerator and denominator. So 5 is a common factor. This can be cancelled. That is

$$\frac{5}{25 + 15x} = \frac{1 \times 5}{5(5 + 3x)} = \frac{1 \times \cancel{5}}{\cancel{5}(5 + 3x)} = \frac{1}{5 + 3x}$$

It is very important to note that the number 5 which has been cancelled is a common factor. It is incorrect to try to cancel terms which are not common factors.

8.2 Simplify $\dfrac{4x}{3x^2 + x}$

Solution Note that the denominator factorizes to $x(3x + 1)$. Once both numerator and denominator have been factorized, any common factors are cancelled. So

$$\frac{4x}{3x^2 + x} = \frac{4x}{x(3x + 1)} = \frac{4\cancel{x}}{\cancel{x}(3x + 1)} = \frac{4}{3x + 1}$$

Note that the factor x is common to both numerator and denominator and so has been cancelled.

8.3 Simplify $\dfrac{x}{x^2 + 2x}$

Solution Note that the denominator factorizes to $x(x + 2)$. Also note that the numerator can be written as $1 \times x$. So

$$\frac{x}{x^2 + 2x} = \frac{1 \times x}{x(x + 2)} = \frac{1 \times \cancel{x}}{\cancel{x}(x + 2)} = \frac{1}{x + 2}$$

8.4 Simplify $\dfrac{x + 2}{x^2 + 3x + 2}$

Solution The denominator is factorized and then any common factors are cancelled:

$$\frac{x + 2}{x^2 + 3x + 2} = \frac{1(x + 2)}{(x + 2)(x + 1)} = \frac{1}{x + 1}$$

8.5 Simplify

(a) $\dfrac{3x + xy}{x^2 + 5x}$ (b) $\dfrac{x^2 - 1}{x^2 + 3x + 2}$

Solution   The numerator and denominator are both factorized and any common factors are cancelled.

(a)

$$\frac{3x + xy}{x^2 + 5x} = \frac{x(3 + y)}{x^2 + 5x} = \frac{\cancel{x}(3 + y)}{\cancel{x}(x + 5)} = \frac{3 + y}{5 + x}$$

(b)

$$\frac{x^2 - 1}{x^2 + 3x + 2} = \frac{(x + 1)(x - 1)}{(x + 1)(x + 2)} = \frac{x - 1}{x + 2}$$

Exercise 8.2

1. Simplify the following algebraic fractions:

   (a) $\dfrac{4}{12 + 8x}$ (b) $\dfrac{5 + 10x}{5}$ (c) $\dfrac{2}{4 + 14x}$

   (d) $\dfrac{2x}{4 + 14x}$ (e) $\dfrac{2x}{2 + 14x}$

2. Simplify the following:

   (a) $\dfrac{5}{15 + 10x}$ (b) $\dfrac{2x}{x^2 + 7x}$ (c) $\dfrac{2x + 8}{x^2 + 2x - 8}$

   (d) $\dfrac{7ab}{a^2b^2 + 9ab}$ (e) $\dfrac{xy}{xy + x}$

3. Simplify

   (a) $\dfrac{x - 4}{(x - 4)(x - 2)}$ (b) $\dfrac{2x - 4}{x^2 + x - 6}$

   (c) $\dfrac{3x}{3x^2 + 6x}$ (d) $\dfrac{x^2 + 2x + 1}{x^2 - 2x - 3}$

8.3   Multiplication and division of algebraic fractions

To multiply two algebraic fractions together we multiply their numerators together and multiply their denominators together:

KEY POINT

$$\frac{a}{b} \times \frac{c}{d} = \frac{a \times c}{b \times d}$$

Any common factors in the result should be cancelled.

Worked examples

8.6 Simplify

$$\frac{4}{5} \times \frac{x}{y}$$

Solution We multiply the numerators together and multiply the denominators together. That is,

$$\frac{4}{5} \times \frac{x}{y} = \frac{4x}{5y}$$

8.7 Simplify

$$\frac{4}{x} \times \frac{3y}{16}$$

Solution The numerators are multiplied together and the denominators are multiplied together. Therefore,

$$\frac{4}{x} \times \frac{3y}{16} = \frac{4 \times 3y}{16x}$$

Because $16x = 4 \times 4x$, the common factor 4 can be cancelled. So

$$\frac{4 \times 3y}{16x} = \frac{\cancel{4} \times 3y}{\cancel{4} \times 4x} = \frac{3y}{4x}$$

8.8 Simplify

$$\text{(a)} \ \frac{1}{2} \times x \ \text{(b)} \ \frac{1}{2} \times (a + b)$$

Solution (a) Writing x as $x/1$ we can state

$$\frac{1}{2} \times x = \frac{1}{2} \times \frac{x}{1} = \frac{1 \times x}{2 \times 1} = \frac{x}{2}$$

(b) Writing $a + b$ as $\dfrac{a+b}{1}$ we can state

$$\frac{1}{2} \times (a + b) = \frac{1}{2} \times \frac{(a+b)}{1} = \frac{1 \times (a+b)}{2 \times 1} = \frac{a+b}{2}$$

8.9 Simplify

$$\frac{4x^2}{y} \times \frac{3x^3}{yz}$$

Solution We multiply the numerators together and multiply the denominators together.

$$\frac{4x^2}{y} \times \frac{3x^3}{yz} = \frac{4x^2 \times 3x^3}{y \times yz} = \frac{12x^5}{y^2z}$$

8.10 Simplify

$$5 \times \left(\frac{x-3}{25} \right)$$

Solution This means

$$\frac{5}{1} \times \frac{x-3}{25}$$

which equals

$$\frac{5 \times (x-3)}{1 \times 25}$$

A common factor of 5 can be cancelled from the numerator and denominator to give

$$\frac{(x-3)}{5}$$

8.11 Simplify

$$-\frac{1}{5} \times \frac{3x-4}{8}$$

Solution We can write

$$-\frac{1}{5} \times \frac{3x-4}{8} = -\frac{1 \times (3x-4)}{5 \times 8} = -\frac{3x-4}{40}$$

Note that the answer can also be expressed as $(4-3x)/40$ because

$$-\frac{3x-4}{40} = \frac{-1}{1} \times \frac{3x-4}{40} = \frac{-3x+4}{40} = \frac{4-3x}{40}$$

You should be aware from the last example that a solution can often be expressed in a number of equivalent ways.

Worked examples

8.12 Simplify

$$\frac{a}{a+b} \times \frac{b}{5a^2}$$

Solution

$$\frac{a}{a+b} \times \frac{b}{5a^2} = \frac{ab}{5a^2(a+b)}$$

Cancelling the common factor of a in numerator and denominator gives

$$\frac{b}{5a(a+b)}$$

8.13   Simplify

$$\frac{x^2 + 4x + 3}{2x + 8} \times \frac{x + 4}{x + 1}$$

Solution   Before multiplying the two fractions together we should try to factorize if possible so that common factors can be identified. By factorizing, we can write the given expressions as

$$\frac{(x + 1)(x + 3)}{2(x + 4)} \times \frac{x + 4}{x + 1} = \frac{(x + 1)(x + 3)(x + 4)}{2(x + 4)(x + 1)}$$

Cancelling common factors this simplifies to just

$$\frac{x + 3}{2}$$

Division is performed by inverting the second fraction and multiplying.

KEY POINT

$$\frac{a}{b} \div \frac{c}{d} = \frac{a}{b} \times \frac{d}{c}$$

Worked examples

8.14   Simplify

$$\frac{10a}{b} \div \frac{a^2}{3b}$$

Solution   The second fraction is inverted and then multiplied by the first, that is,

$$\frac{10a}{b} \div \frac{a^2}{3b} = \frac{10a}{b} \times \frac{3b}{a^2} = \frac{30ab}{a^2b} = \frac{30}{a}$$

8.15   Simplify

$$\frac{x^2 y^3}{z} \div \frac{y}{x}$$

Solution

$$\frac{x^2 y^3}{z} \div \frac{y}{x} = \frac{x^2 y^3}{z} \times \frac{x}{y} = \frac{x^3 y^3}{zy}$$

Any common factors in the result can be cancelled. So

$$\frac{x^3 y^3}{zy} = \frac{x^3 y^2}{z}$$

Self-assessment questions 8.3

1. The technique of multiplying and dividing algebraic fractions is identical to that used for numbers – true or false?

Exercise 8.3

1. Simplify the following:

(a) $\dfrac{5}{4} \times \dfrac{a}{25}$ (b) $\dfrac{5}{4} \times \dfrac{a}{b}$ (c) $\dfrac{8a}{b^2} \times \dfrac{b}{16a^2}$

(d) $\dfrac{9x}{3y} \times \dfrac{2x}{y^2}$ (e) $\dfrac{3}{5a} \times \dfrac{b}{a}$

(f) $\dfrac{1}{4} \times \dfrac{x}{y}$ (g) $\dfrac{1}{3} \times \dfrac{x}{x+y}$

(h) $\dfrac{x-3}{x+4} \times \dfrac{1}{3x-9}$

2. Simplify the following:

(a) $\dfrac{3}{x} \times \dfrac{xy}{z^3}$ (b) $\dfrac{(3+x)}{x} \div \dfrac{y}{x}$ (c) $\dfrac{4}{3} \div \dfrac{16}{x}$

(d) $\dfrac{a}{bc^2} \times \dfrac{b^2c}{a}$

3. Simplify

(a) $\dfrac{x+2}{(x+5)(x+4)} \times \dfrac{x+5}{x+2}$

(b) $\dfrac{x-2}{4} \div \dfrac{x}{16}$ (c) $\dfrac{12ab}{5ef} \div \dfrac{4ab^2}{f}$

(d) $\dfrac{x+3y}{2x} \div \dfrac{y}{4x^2}$ (e) $\dfrac{3}{x} \times \dfrac{3}{y} \times \dfrac{1}{z}$

| 8.4 | **Addition and subtraction of algebraic fractions** |

For revision of adding and subtracting fractions see §2.3

The method is the same as that for adding or subtracting numerical fractions. Note that it is not correct to simply add or subtract the numerator and denominator. The lowest common denominator must first be found. This is the simplest expression which contains all original denominators as its factors. Each fraction is then written with this common denominator. The fractions can then be added or subtracted by adding or subtracting just the numerators. Consider the following examples.

Worked examples

8.16   Simplify

$$\frac{3}{x} + \frac{4}{x^2}$$

Solution The expression $3/x$ is rewritten as $3x/x^2$ which makes the denominators of both terms x^2 but leaves the value of the expression unaltered. Note that both the original denominators, x and x^2, are factors of the new denominator. We call this denominator the **lowest common denominator.** The fractions are then added by adding just the numerators. That is,

$$\frac{3x}{x^2} + \frac{4}{x^2} = \frac{3x+4}{x^2}$$

8.17 Express $\dfrac{5}{a} - \dfrac{4}{b}$ as a single fraction.

Solution Both fractions are rewritten to have the same denominator. The simplest expression containing both a and b as its factors is ab. Therefore ab is the lowest common denominator. Then

$$\frac{5}{a} - \frac{4}{b} = \frac{5b}{ab} - \frac{4a}{ab} = \frac{5b-4a}{ab}$$

8.10 Write $\dfrac{4}{x+y} - \dfrac{3}{y}$ as a single fraction.

Solution The simplest expression which contains both denominators as its factors is $(x+y)y$. We must rewrite each term so that it has this denominator.

$$\frac{4}{x+y} = \frac{4}{x+y} \times \frac{y}{y} = \frac{4y}{(x+y)y}$$

Similarly,

$$\frac{3}{y} = \frac{3}{y} \times \frac{x+y}{x+y} = \frac{3(x+y)}{(x+y)y}$$

The fractions are then subtracted by subtracting just the numerators:

$$\frac{4}{x+y} - \frac{3}{y} = \frac{4y}{(x+y)y} - \frac{3(x+y)}{(x+y)y} = \frac{4y-3(x+y)}{(x+y)y}$$

which simplifies to

$$\frac{y-3x}{(x+y)y}$$

Self-assessment questions 8.4

1. Explain what is meant by the 'lowest common denominator' and how it is found.

Exercise 8.4

1. Express each of the following as a single fraction:

(a) $\dfrac{3}{x} + \dfrac{4}{y}$ (b) $\dfrac{3}{x^2} + \dfrac{4y}{x}$ (c) $\dfrac{4ab}{x} + \dfrac{3ab}{2y}$

(d) $\dfrac{4xy}{a} + \dfrac{3xy}{2b}$ (e) $\dfrac{3}{x} - \dfrac{6}{2x}$ (f) $\dfrac{3x}{2y} - \dfrac{7y}{4x}$

(g) $\dfrac{3}{x+y} - \dfrac{2}{y}$ (h) $\dfrac{1}{a+b} - \dfrac{1}{a-b}$

(i) $2x + \dfrac{1}{2x}$ (j) $2x - \dfrac{1}{2x}$

2. Express each of the following as a single fraction:

(a) $\dfrac{x}{y} + \dfrac{3x^2}{z}$ (b) $\dfrac{4}{a} + \dfrac{5}{b}$ (c) $\dfrac{6x}{y} - \dfrac{2y}{x}$

(d) $3x - \dfrac{3x+1}{4}$ (e) $\dfrac{5a}{12} + \dfrac{9a}{18}$ (f) $\dfrac{x-3}{4} + \dfrac{3}{5}$

Test and assignment exercises 8

1. Simplify

(a) $\dfrac{5a}{4a+3ab}$ (b) $\dfrac{5a}{30a+15b}$ (c) $\dfrac{5ab}{30a+15b}$

(d) $\dfrac{5ab}{ab+7ab}$ (e) $\dfrac{y}{13y+y^2}$ (f) $\dfrac{13y+y^2}{y}$

2. Simplify the following:

(a) $\dfrac{5a}{7} \times \dfrac{14b}{2}$ (b) $\dfrac{3}{x} + \dfrac{7}{3x}$ (c) $t - \dfrac{4-t}{2}$

(d) $4(x+3) - \dfrac{(4x-5)}{3}$ (e) $\dfrac{7}{x} + \dfrac{3}{2x} + \dfrac{5}{3x}$

(f) $x + \dfrac{3x}{y}$ (g) $xy + \dfrac{1}{xy}$

(h) $\dfrac{1}{x+y} + \dfrac{2}{x-y}$

3. Simplify the following:

(a) $\dfrac{y}{9} + \dfrac{2y}{7}$ (b) $\dfrac{3}{x} - \dfrac{5}{3x} + \dfrac{4}{5x}$ (c) $\dfrac{3x}{2y} + \dfrac{5y}{6x}$

(d) $m + \dfrac{m+n}{2}$ (e) $m - \dfrac{m+n}{2}$

(f) $m - \dfrac{m-n}{2}$ (g) $\dfrac{3s-5}{10} - \dfrac{2s-3}{15}$

4. Simplify

$$\dfrac{x^2 - x}{x - 1}$$

9 Transposing formulae

Objectives

This chapter

- explains how formulae can be rearranged or transposed

9.1 Rearranging a formula

In the formula for the area of a circle, $A = \pi r^2$, we say that A is the **subject** of the formula. The subject appears by itself on one side, usually the left, of the formula, and nowhere else. If we are asked to **transpose** the formula for r, then we must rearrange the formula so that r becomes the subject. The rules for transposing formulae are quite simple. Essentially whatever you do to one side you must also do the same to the whole of the other side.

KEY POINT

To transpose a formula you may:

- add the same quantity to both sides
- subtract the same quantity from both sides
- multiply or divide both sides by the same quantity
- perform operations on both sides, such as 'square both sides', 'square root both sides' etc.

Worked examples

9.1 The circumference C of a circle is given by the formula $C = 2\pi r$. Transpose this to make r the subject.

Solution   The intention is to obtain r by itself on the left-hand side. Starting with $C = 2\pi r$ we must try to isolate r. Dividing both sides by 2π we find

$$C = 2\pi r$$

$$\frac{C}{2\pi} = \frac{2\pi r}{2\pi}$$

$$\frac{C}{2\pi} = r \quad \text{by cancelling the common factor } 2\pi \text{ on the right}$$

Finally we can write $r = C/2\pi$ and the formula has been transposed for r.

9.2   Transpose the formula $y = 3(x + 7)$ for x.

Solution   We must try to obtain x on its own on the left-hand side. This can be done in stages. Dividing both sides by 3 we find

$$\frac{y}{3} = \frac{3(x + 7)}{3} = x + 7 \qquad \text{by cancelling the common factor } 3$$

Subtracting 7 from both sides then gives

$$\frac{y}{3} - 7 = x + 7 - 7$$

so that

$$\frac{y}{3} - 7 = x$$

Equivalently we can write $x = \dfrac{y}{3} - 7$, and the formula has been transposed for x.

Alternatively, again starting from $y = 3(x + 7)$ we could proceed as follows. Removing the brackets we obtain $y = 3x + 21$. Then, subtracting 21 from both sides,

$$y - 21 = 3x + 21 - 21 = 3x$$

Dividing both sides by 3 will give x on its own:

$$\frac{y - 21}{3} = x \qquad \text{so that} \qquad x = \frac{y - 21}{3}$$

Noting that

$$\frac{y - 21}{3} = \frac{y}{3} - \frac{21}{3} = \frac{y}{3} - 7$$

we see that this answer is equivalent to the expression obtained previously.

9.3 Transpose the formula $y - z = 3(x + 2)$ for x.

Solution Dividing both sides by 3 we find

$$\frac{y - z}{3} = x + 2$$

Subtracting 2 from both sides gives

$$\frac{y - z}{3} - 2 = x + 2 - 2$$

$$= x$$

so that

$$x = \frac{y - z}{3} - 2$$

9.4 Transpose $x + xy = 7$, (a) for x, (b) for y.

Solution (a) In the expression $x + xy$, note that x is a common factor. We can then write the given formula as $x(1 + y) = 7$. Dividing both sides by $1 + y$ will isolate x. That is,

$$\frac{x(1 + y)}{1 + y} = \frac{7}{1 + y}$$

Cancelling the common factor $1 + y$ on the left-hand side gives

$$x = \frac{7}{1 + y}$$

and the formula has been transposed for x.

(b) To transpose $x + xy = 7$ for y we first subtract x from both sides to give $xy = 7 - x$. Finally dividing both sides by x gives

$$\frac{xy}{x} = \frac{7 - x}{x}$$

Cancelling the common factor of x on the left-hand side gives

$$y = \frac{7 - x}{x}$$ which is the required transposition.

9.5 Transpose the formula $I = V/R$ for R.

Solution Starting with $I = V/R$ we first multiply both sides by R. This gives

$$IR = \frac{V}{R} \times R = V$$

Then, dividing both sides by I gives

$$\frac{IR}{I} = \frac{V}{I}$$

so that $R = V/I$.

9.6 Make x the subject of the formula $y = \dfrac{4}{x-7}$

Solution We must try to obtain x on its own on the left-hand side. It is often useful to multiply both sides of the formula by the same quantity in order to remove fractions. Multiplying both sides by $(x-7)$ we find

$$(x-7)y = (x-7) \times \frac{4}{x-7} = 4$$

Removing the brackets on the left-hand side we find

$$xy - 7y = 4 \qquad \text{so that} \qquad xy = 7y + 4$$

Finally, dividing both sides by y we obtain

$$\frac{xy}{y} = \frac{7y+4}{y}$$

that is,

$$x = \frac{7y+4}{y}$$

9.7 Transpose the formula

$$T = 2\pi \sqrt{\frac{l}{g}}$$

to make l the subject.

Solution We must attempt to isolate l. We can do this in stages as follows. Dividing both sides by 2π we find

$$\frac{T}{2\pi} = \frac{2\pi\sqrt{l/g}}{2\pi} = \sqrt{\frac{l}{g}} \qquad \text{by cancelling the common factor } 2\pi$$

Recall from §5.3 that $(\sqrt{a})^2 = a$

The square root sign over the l/g can be removed by squaring both sides. Recall that if a term containing a square root is squared then the square root will disappear.

$$\left(\frac{T}{2\pi}\right)^2 = \frac{l}{g}$$

Then multiplying both sides by g we find

$$g\left(\frac{T}{2\pi}\right)^2 = l$$

Equivalently we have

$$l = g\left(\frac{T}{2\pi}\right)^2$$

and the formula has been transposed for l.

Self-assessment questions 9.1

1 Explain what is meant by the 'subject' of a formula.
2. In what circumstances might you want to transpose a formula?
3. Explain what processes are allowed when transposing a formula.

Exercise 9.1

1. Transpose each of the following formulae to make x the subject:

(a) $y = 3x$ (b) $y = \dfrac{1}{x}$ (c) $y = 7x - 5$

(d) $y = \dfrac{1}{2}x - 7$ (e) $y = \dfrac{1}{2x}$

(f) $y = \dfrac{1}{2x + 1}$ (g) $y = \dfrac{1}{2x} + 1$

(h) $y = 18x - 21$ (i) $y = 19 - 8x$

2. Transpose $y = mx + c$, (a) for m, (b) for x, (c) for c.

3. Transpose the following formulae for x:

(a) $y = 13(x - 2)$ (b) $y = x\left(1 + \dfrac{1}{x}\right)$

(c) $y = a + t(x - 3)$

4. Transpose each of the following formulae to make the given variable the subject:

(a) $y = 7x + 11$, for x (b) $V = IR$, for I

(c) $V = \dfrac{4}{3}\pi r^3$, for r (d) $F = ma$, for m

5. Make n the subject of the formula $l = a + (n - 1)d$.

6. If $m = n + t\sqrt{x}$, find an expression for x.

7. Make x the subject of the following formulae:

(a) $y = 1 - x^2$ (b) $y = \dfrac{1}{1 - x^2}$

(c) $y = \dfrac{1 - x^2}{1 + x^2}$

Test and assignment exercises 9

1. Make x the subject of the following formulae:

(a) $y = 13x - 18$ (b) $y = \dfrac{13}{x + 18}$

(c) $y = \dfrac{13}{x} + 18$

2. Transpose the following formulae:

(a) $y = \dfrac{7 + x}{14}$ for x (b) $E = \dfrac{1}{2}mv^2$ for v

(c) $8x - 13y = 12$ for y (d) $y = \dfrac{x^2}{2g}$ for x

3. Make x the subject of the formula $v = k/\sqrt{x}$.

4. Transpose the formula $V = \pi r^2 h$ for h.

5. Make r the subject of the formula $y = H + Cr$.

6. Transpose the formula $s = ut + \dfrac{1}{2}at^2$ for a.

7. Transpose the formula $v^2 = u^2 + 2as$ for u.

8. Transpose the formula $k = pv^3$ for v.

10 Solving equations

Objectives

This chapter

- explains what is meant by an equation and its solution
- shows how to solve linear, simultaneous and quadratic equations

An **equation** states that two quantities are equal, and will always contain an **unknown quantity** which we wish to find. For example, in the equation $5x + 10 = 20$ the unknown quantity is x. To **solve** an equation means to find all values of the unknown quantity which can be substituted into the equation so that the left side equals the right side. Each such value is called a **solution** or alternatively a **root** of the equation. In the example above, the solution is $x = 2$ because when $x = 2$ is substituted both the left side and the right side equal 20. The value $x = 2$ is said to **satisfy** the equation.

10.1 Solving linear equations

A **linear** equation is one of the form $ax + b = 0$ where a and b are numbers and the unknown quantity is x. The number a is called the **coefficient** of x. The number b is called the **constant term**. For example, $3x + 7 = 0$ is a linear equation. The coefficient of x is 3 and the constant term is 7. Similarly, $-2x + 17.5 = 0$ is a linear equation. The coefficient of x is -2 and the constant term is 17.5. Note that the unknown quantity only occurs to the first power. Linear equations may appear in other forms which may seem to be different but are nevertheless equivalent. Thus

$$4x + 13 = -7, \qquad 3 - 14x = 0 \qquad \text{and} \qquad 3x + 7 = 2x - 4$$

are all linear equations which could be written in the form $ax + b = 0$ if necessary.

An equation such as $3x^2 + 2 = 0$ is not linear because the unknown quantity occurs to the power 2. Linear equations are solved by trying to obtain the unknown quantity on its own on the left-hand side, that is, by making the unknown quantity the subject of the equation. This is done using the same rules given for transposing formulae in Chapter 9. Consider the following examples.

Worked example

10.1   Solve the equation $x + 10 = 0$.

Solution   We make x the subject by subtracting 10 from both sides to give $x = -10$. Therefore $x = -10$ is the solution. It can be easily checked by substituting into the original equation:

$$(-10) + 10 = 0 \text{ as required}$$

Note from the last example that the solution should be checked by substitution to ensure it satisfies the given equation. If it does not then a mistake has been made.

Worked examples

10.2   Solve the equation $4x + 8 = 0$.

Solution   In order to find the unknown quantity x we attempt to make it the subject of the equation. Subtracting 8 from both sides we find

$$4x + 8 - 8 = 0 - 8 = -8$$

that is,

$$4x = -8$$

Then, dividing both sides by 4 gives

$$\frac{4x}{4} = \frac{-8}{4}$$

so that $x = -2$. The solution of the equation $4x + 8 = 0$ is $x = -2$.

10.3   Solve the equation $5x + 17 = 4x - 3$.

Solution   First we collect all terms involving x together. This is done by subtracting $4x$ from both sides to remove this term from the right. This gives

$$5x - 4x + 17 = -3$$

that is, $x + 17 = -3$. To make x the subject of this equation we now subtract 17 from both sides to give $x = -3 - 17 = -20$. The solution of the equation $5x + 17 = 4x - 3$ is $x = -20$. Note that the answer can be easily checked by substituting $x = -20$ into the original equation and verifying that the left side equals the right side.

10.4　Solve the equation $\dfrac{x-3}{4} = 1$.

Solution　We attempt to obtain x on its own. First note that if we multiply both sides of the equation by 4 this will remove the 4 in the denominator. That is,

$$4 \times \left(\frac{x-3}{4} \right) = 4 \times 1$$

so that

$$x - 3 = 4$$

Finally adding 3 to both sides gives $x = 7$.

Self-assessment questions 10.1

1. Explain what is meant by a root of an equation.
2. Explain what is meant by a linear equation.
3. State the rules which can be used to solve a linear equation.
4. You may think that a formula and an equation look very similar. Try to explain the distinction between a formula and an equation.

Exercise 10.1

1. Solve the following linear equations:

   (a) $3x = 9$ (b) $\dfrac{x}{3} = 9$ (c) $3t + 6 = 0$

   (d) $3x - 13 = 2x + 9$ (e) $3x + 17 = 21$

   (f) $4x - 20 = 3x + 16$

   (g) $5 - 2x = 2 + 3x$

   (h) $\dfrac{x+3}{2} = 3$ (i) $\dfrac{3x+2}{2} + 3x = 1$

2. Solve the following equations:

   (a) $5(x + 2) = 13$
   (b) $3(x - 7) = 2(x + 1)$
   (c) $5(1 - 2x) = 2(4 - 2x)$

3. Solve the following equations:

   (a) $3t + 7 = 4t - 2$ (b) $3v = 17 - 4v$
   (c) $3s + 2 = 14(s - 1)$

10.2　**Solving simultaneous equations**

Sometimes equations contain more than one unknown quantity. When this happens there are usually two or more equations. For example in the two equations

$$x + 2y = 14 \qquad 3x + y = 17$$

the unknowns are x and y. Such equations are called **simultaneous equations** and to solve them we must find values of x and y which satisfy both equations at the same time. If we substitute $x = 4$ and $y = 5$ into either of the two equations above we see that the equation is satisfied. We shall demonstrate how simultaneous equations can be solved by removing, or **eliminating**, one of the unknowns.

Worked examples

10.5 Solve the simultaneous equations

$$x + 3y = 14 \tag{10.1}$$

$$2x - 3y = -8 \tag{10.2}$$

Solution Note that if these two equations are added the unknown y is removed or eliminated:

$$\begin{array}{r} x + 3y = 14 \\ 2x - 3y = -8 \end{array} +$$

$$\overline{3x \qquad = 6}$$

so that $x = 2$. To find y we substitute $x = 2$ into either equation. Substituting into Equation 10.1 gives

$$2 + 3y = 14$$

Solving this linear equation will give y. We have

$$2 + 3y = 14$$

$$3y = 14 - 2 = 12$$

$$y = \frac{12}{3} = 4$$

Therefore the solution of the simultaneous equations is $x = 2$ and $y = 4$. Note that these solutions should be checked by substituting back into both given equations to check that the left-hand side equals the right-hand side.

10.6 Solve the simultaneous equations

$$5x + 4y = 7 \tag{10.3}$$

$$3x - y = 11 \tag{10.4}$$

Solution Note that in this example, if we multiply the first equation by 3 and the second by 5 we shall have the same coefficient of x in both equations. This gives

$$15x + 12y = 21 \tag{10.5}$$

$$15x - 5y = 55 \tag{10.6}$$

We can now eliminate x by subtracting Equation 10.6 from Equation 10.5, giving

$$\begin{array}{r} 15x + 12y = 21 \\ 15x - 5y = 55 \end{array} -$$

$$\overline{17y = -34}$$

from which $y = -2$. In order to find x we substitute our solution for y

into either of the given equations. Substituting into Equation 10.3 gives

$$5x + 4(-2) = 7 \qquad \text{so that} \qquad 5x = 15 \qquad \text{or} \qquad x = 3$$

The solution of the simultaneous equations is therefore $x = 3$, $y = -2$.

Exercise 10.2

1. Solve the following pairs of simultaneous equations:

   (a) $3x + y = 1$, $2x - y = 2$
   (b) $4x + 5y = 21$, $3x + 5y = 17$

   (c) $2x - y = 17$, $x + 3y = 12$
   (d) $-2x + y = -21$, $x + 3y = -14$
   (e) $-x + y = -10$, $3x + 7y = 20$
   (f) $4x - 2y = 2$, $3x - y = 4$

10.3   Solving quadratic equations

A **quadratic equation** is an equation of the form $ax^2 + bx + c = 0$ where a, b and c are numbers and x is the unknown quantity we wish to find. The number a is the coefficient of x^2, b is the coefficient of x and c is the constant term. Sometimes b or c may be zero, although a can never be zero. For example

$$x^2 + 7x + 2 = 0 \qquad 3x^2 - 2 = 0 \qquad -2x^2 + 3x = 0 \qquad 8x^2 = 0$$

are all quadratic equations.

KEY POINT

> A quadratic equation has the form $ax^2 + bx + c = 0$ where a, b and c are numbers, and x represents the unknown we wish to find.

Worked examples

10.7 State the coefficient of x^2 and the coefficient of x in the following quadratic equations:

(a) $4x^2 + 3x - 2 = 0$ (b) $x^2 - 23x + 17 = 0$ (c) $-x^2 + 19 = 0$

Solution   (a) In the equation $4x^2 + 3x - 2 = 0$ the coefficient of x^2 is 4 and the coefficient of x is 3.

(b) In the equation $x^2 - 23x + 17 = 0$ the coefficient of x^2 is 1 and the coefficient of x is -23.

(c) In the equation $-x^2 + 19 = 0$ the coefficient of x^2 is -1 and the coefficient of x is zero, since there is no term involving just x.

Solution by factorization

If the quadratic expression on the left-hand side of the equation can be factorized solutions can be found using the method in the following examples.

Worked examples

10.8 Solve the quadratic equation $x^2 + 3x - 10 = 0$.

Solution The left-hand side of the equation can be factorized to give

$$x^2 + 3x - 10 = (x + 5)(x - 2) = 0$$

Whenever the product of two quantities equals zero, then one or both of these quantities must be zero. It follows that either $x + 5 = 0$ or $x - 2 = 0$ from which $x = -5$ and $x = 2$ are the required solutions.

10.9 Solve the quadratic equation $6x^2 + 5x - 4 = 0$.

Solution The left-hand side of the equation can be factorized to give

$$6x^2 + 5x - 4 = (2x - 1)(3x + 4) = 0$$

It follows that either $2x - 1 = 0$ or $3x + 4 = 0$ from which $x = \frac{1}{2}$ and $x = -\frac{4}{3}$ are the required solutions.

10.10 Solve the quadratic equation $x^2 - 8x = 0$.

Solution Factorizing the left-hand side gives

$$x^2 - 8x = x(x - 8) = 0$$

from which either $x = 0$ or $x - 8 = 0$. The solutions are therefore $x = 0$ and $x = 8$.

10.11 Solve the quadratic equation $x^2 - 36 = 0$.

Solution The left-hand side factorizes to give

$$x^2 - 36 = (x + 6)(x - 6) = 0$$

so that $x + 6 = 0$ or $x - 6 = 0$. The solutions are therefore $x = -6$ and $x = 6$.

10.12 Solve the equation $x^2 = 81$.

Solution Writing this in the standard form of a quadratic equation we obtain $x^2 - 81 = 0$. The left-hand side can be factorized to give

$$x^2 - 81 = (x - 9)(x + 9) = 0$$

from which $x - 9 = 0$ or $x + 9 = 0$. The solutions are then $x = 9$ and $x = -9$.

Solution of quadratic equations using the formula

It should be noted that not all quadratic equations have solutions. However, when it is difficult or impossible to factorize the quadratic expression $ax^2 + bx + c$, solutions of a quadratic equation can be sought using the following formula:

KEY POINT

$$x = \frac{-b \pm \sqrt{b^2 - 4ac}}{2a}$$

This formula gives possibly two solutions: one solution is obtained by taking the positive square root and the second solution by taking the negative square root.

Worked examples

10.13   Solve the equation $x^2 + 9x + 20 = 0$ using the formula.

Solution   Comparing the given equation with the standard form $ax^2 + bx + c = 0$ we see that $a = 1$, $b = 9$ and $c = 20$. These values are substituted into the formula

$$
\begin{aligned}
x &= \frac{-b \pm \sqrt{b^2 - 4ac}}{2a} \\
&= \frac{-9 \pm \sqrt{81 - 4(1)(20)}}{(2)(1)} \\
&= \frac{-9 \pm \sqrt{81 - 80}}{2} \\
&= \frac{-9 \pm \sqrt{1}}{2} \\
&= \frac{-9 \pm 1}{2} \\
&= \begin{cases} -4 & \text{by taking the } +\text{ve square root} \\ -5 & \text{by taking the } -\text{ve square root} \end{cases}
\end{aligned}
$$

The two solutions are therefore $x = -4$ and $x = -5$.

10.14   Solve the equation $2x^2 - 3x - 7 = 0$ using the formula.

Solution   In this example $a = 2$, $b = -3$ and $c = -7$. Care should be taken with the negative signs. Substituting these into the formula we find

$$x = \frac{-(-3) \pm \sqrt{(-3)^2 - 4(2)(-7)}}{2(2)}$$

$$= \frac{3 \pm \sqrt{9 + 56}}{4}$$

$$= \frac{3 \pm \sqrt{65}}{4}$$

$$= \frac{3 \pm 8.062}{4}$$

$$= \begin{cases} 2.766 & \text{by taking the +ve square root} \\ -1.266 & \text{by taking the -ve square root} \end{cases}$$

The two solutions are therefore $x = 2.766$ and $x = -1.266$.

If the values of a, b and c are such that $b^2 - 4ac$ is positive the formula will produce two solutions known as **distinct real roots** of the equation. If $b^2 - 4ac = 0$ there will be a single root known as a **repeated root**. Some books refer to the equation having **equal roots**. If the equation is such that $b^2 - 4ac$ is negative the formula requires us to find the square root of a negative number. In ordinary arithmetic this is impossible and we say that in such a case the quadratic equation does not possess any real roots.

Worked examples

10.15   Can the equation $x^2 + 49 = 0$ be solved using the formula?

Solution   In this example $a = 1$, $b = 0$ and $c = 49$. Therefore $b^2 - 4ac = 0 - 4(1)(49)$ which is negative. Therefore this equation does not possess any real roots.

10.16   Use the formula to solve the equation $4x^2 + 4x + 1 = 0$.

Solution   In this example $a = 4$, $b = 4$ and $c = 1$. Applying the formula gives

$$x = \frac{-4 \pm \sqrt{4^2 - 4(4)(1)}}{(2)(4)}$$

$$= \frac{-4 \pm \sqrt{16 - 16}}{8}$$

$$= \frac{-4 \pm 0}{8} = -\frac{1}{2}$$

There is a single, repeated root $x = -\frac{1}{2}$. Note that $b^2 - 4ac = 0$.

Self-assessment questions 10.3

1. Under what conditions will a quadratic equation possess distinct real roots?
2. Under what conditions will a quadratic equation possess a repeated root?

Exercise 10.3

1. Solve the following quadratic equations by factorization:

   (a) $x^2 + x - 2 = 0$ (b) $x^2 - 8x + 15 = 0$
   (c) $4x^2 + 6x + 2 = 0$ (d) $x^2 - 6x + 9 = 0$
   (e) $x^2 - 81 = 0$

2. Solve the following quadratic equations using the formula:

   (a) $3x^2 - 6x - 5 = 0$
   (b) $x^2 + 3x - 77 = 0$ (c) $2x^2 - 9x + 2 = 0$

3. Solve the following quadratic equations:

   (a) $6x^2 + 13x + 6 = 0$
   (b) $3t^2 + 13t + 12 = 0$ (c) $t^2 - 7t + 3 = 0$

Test and assignment exercises 10

1. Solve the following equations:

   (a) $13t - 7 = 2t + 5$
   (b) $-3t + 9 = 13 - 7t$ (c) $-5t = 0$

2. Solve the following equations:

   (a) $4x = 16$ (b) $\dfrac{x}{12} = 9$ (c) $4x - 13 = 3$

   (d) $4x - 14 = 2x + 8$ (e) $3x - 17 = 4$

   (f) $7 - x = 9 + 3x$ (g) $4(2 - x) = 8$

3. Solve the following equations:

   (a) $2y = 8$ (b) $5u = 14u + 3$ (c) $5 = 4I$

   (d) $13i + 7 = 2i - 9$ (e) $\dfrac{1}{4}x + 9 = 3 - \dfrac{1}{2}x$

   (f) $\dfrac{4x + 2}{2} + 8x = 0$

4. Solve the following equations:

   (a) $3x^2 - 27 = 0$ (b) $2x^2 + 18x + 14 = 0$
   (c) $x^2 = 16$ (d) $x^2 - x - 72 = 0$
   (e) $2x^2 - 3x - 44 = 0$
   (f) $x^2 - 4x - 21 = 0$

5. The solutions of the equation
   $x^2 + 4x = 0$ are identical to the solutions
   of $3x^2 + 12x = 0$ – true or false?

6. Solve the following simultaneous equations:

   (a) $x - 2y = -11$, $7x + y = -32$
   (b) $2x - y = 2$, $x + 3y = 29$
   (c) $x + y = 19$, $-x + y = 1$

11 Functions

Objectives

This chapter

- explains what is meant by a function
- describes the notation used to write functions
- explains the terms 'independent variable' and 'dependent variable'
- explains what is meant by the inverse of a function

11.1   Definition of a function

A **function** is a rule which receives an input and produces an output. It is shown schematically in Figure 11.1. For example, the rule may be 'add 2 to the input'. If 6 is the input, then $6 + 2 = 8$ will be the output. If -5 is

Figure 11.1
A function produces an output from an input

the input then $-5 + 2 = -3$ will be the output. In general, if x is the input then $x + 2$ will be the output. Figure 11.2 illustrates this function schematically.

Figure 11.2
The function adds 2 to the input

$$x \longrightarrow \boxed{\begin{array}{c}\text{Add 2 to}\\\text{the input}\end{array}} \longrightarrow x + 2$$

For a rule to be a function then it is crucial that only **one** output is produced for any given input.

KEY POINT

A function is a rule which produces a *single* output for any given input.

The input to a function can usually take many values and so is called a **variable**. The output, too, varies depending upon the value of the input, and so is also a variable. The input is referred to as the **independent variable** because we are free to choose its value. The output is called the **dependent variable** because its value depends upon the value of the input.

11.2   Notation used for functions

We usually denote the input, the output and the function by letters or symbols. Commonly we use x to represent the input, y the output and f the function, although other letters will be used as well.

Consider again the example from §11.1. We let f be the function 'add 2 to the input' and we let x be the input. In mathematical notation we write

$$f : x \rightarrow x + 2$$

This means that the function f takes an input x and produces an output $x + 2$.

An alternative, but commonly used, notation is

$$f(x) = x + 2$$

The quantity $f(x)$ does not mean f times x but rather indicates that the function f acts on the quantity in the brackets. Because we also call the output y we can write $y = f(x) = x + 2$, or simply $y = x + 2$.

We could represent the same function using different letters. If h represents the function and t the input then we could write

$$h(t) = t + 2$$

Worked examples

11.1   A function multiplies the input by 4. Write down the function in mathematical notation.

Solution   Let us call the function f and the input x. Then we have

$$f : x \rightarrow 4x \quad \text{or alternatively} \quad f(x) = 4x$$

If we call the output y, we can write $y = f(x) = 4x$, or simply $y = 4x$.

11.2   A function divides the input by 6 and then adds 3 to the result. Write the function in mathematical notation.

Solution   Let us call the function z and the input t. Then we have

$$z(t) = \frac{t}{6} + 3$$

11.3   A function f is given by the rule $f: x \rightarrow 9$, or alternatively as $f(x) = 9$. Describe in words what this function does.

Solution   Whatever the value of the input to this function, the output is always 9.

11.4   A function squares the input and then multiplies the result by 6. Write down the function using mathematical notation.

Solution   Let us call the function f, the input x and the output y. Then

$$y = f(x) = 6x^2$$

11.5   Describe in words what the following functions do:

(a) $h(x) = \dfrac{1}{x}$ (b) $g(t) = t + t^2$

Solution   (a) The function $h(x) = 1/x$ divides 1 by the input.
(b) The function $g(t) = t + t^2$ adds the input to the square of the input.

Often we are given a function and need to calculate the output from a given input.

Worked examples

11.6   A function f is defined by $f(x) = 3x + 1$. Calculate the output when the input is (a) 4, (b) -1, (c) 0.

Solution   The function f multiplies the input by 3 and then adds 1 to the result.

(a) When the input is 4, the output is $3 \times 4 + 1 = 12 + 1 = 13$. We write

$$f(x = 4) = 3(4) + 1 = 12 + 1 = 13$$

or more simply

$$f(4) = 13$$

Note that 4 has been substituted for x in the formula for f.

(b) We require the output when the input is -1, that is, $f(-1)$.

$$f(x = -1) = f(-1) = 3(-1) + 1 = -3 + 1 = -2$$

The output is -2 when the input is -1.

(c) We require the output when the input is 0, that is, $f(0)$.

$$f(x = 0) = f(0) = 3(0) + 1 = 0 + 1 = 1$$

11.7 A function g is defined by $g(t) = 2t^2 - 1$. Find

(a) $g(3)$ (b) $g(0.5)$ (c) $g(-2)$

Solution (a) We obtain $g(3)$ by substituting 3 for t.

$$g(3) = 2(3)^2 - 1 = 2(9) - 1 = 17$$

(b) $g(0.5) = 2(0.5)^2 - 1 = 0.5 - 1 = -0.5$.
(c) $g(-2) = 2(-2)^2 - 1 = 8 - 1 = 7$.

11.8 A function h is defined by $h(x) = \dfrac{x}{3} + 1$. Find

(a) $h(3)$ (b) $h(t)$ (c) $h(\alpha)$ (d) $h(2\alpha)$ (e) $h(2x)$

Solution (a) If $h(x) = \dfrac{x}{3} + 1$ then $h(3) = \dfrac{3}{3} + 1 = 1 + 1 = 2$.

(b) The function h divides the input by 3 and then adds 1. We require the output when the input is t, that is, we require $h(t)$. Now

$$h(t) = \frac{t}{3} + 1$$

since the input t has been divided by 3 and then 1 has been added to the result. Note that $h(t)$ is obtained by substituting t in place of x in $h(x)$.

(c) We require the output when the input is α. This is obtained by substituting α in place of x. We find

$$h(\alpha) = \frac{\alpha}{3} + 1$$

(d) We require the output when the input is 2α. We substitute 2α in place of x. This gives

$$h(2\alpha) = \frac{2\alpha}{3} + 1$$

(e) We require the output when the input is $2x$. We substitute $2x$ in place of x. That is,

$$h(2x) = \frac{2x}{3} + 1$$

11.9 Given $f(x) = x^2 + x - 1$ write expressions for

(a) $f(\alpha)$ (b) $f(x + 1)$ (c) $f(2t)$

Solution (a) Substituting α in place of x we obtain

$$f(\alpha) = \alpha^2 + \alpha - 1$$

(b) Substituting $x + 1$ for x we obtain

$$\begin{aligned} f(x+1) &= (x+1)^2 + (x+1) - 1 \\ &= x^2 + 2x + 1 + x + 1 - 1 \\ &= x^2 + 3x + 1 \end{aligned}$$

(c) Substituting $2t$ in place of x we obtain

$$f(2t) = (2t)^2 + 2t - 1 = 4t^2 + 2t - 1$$

Sometimes a function uses different rules on different intervals. For example we could define a function as

$$f(x) = \begin{cases} 3x & \text{when} & 0 \le x \le 4 \\ 2x + 6 & \text{when} & 4 < x < 5 \\ 9 & \text{when} & x \ge 5 \end{cases}$$

Here the function is defined in three 'pieces'. The value of x determines which part of the definition is used to evaluate the function.

Worked examples

11.10 A function is defined by

$$y(x) = \begin{cases} x^2 + 1 & \text{when} & -1 \le x \le 2 \\ 3x & \text{when} & 2 < x \le 6 \\ 2x + 1 & \text{when} & x > 6 \end{cases}$$

Evaluate   (a) $y(0)$ (b) $y(4)$ (c) $y(2)$ (d) $y(7)$

Solution (a) We require the value of y when $x = 0$. Since 0 lies between -1 and 2 we use the first part of the definition, that is, $y = x^2 + 1$. Hence

$$y(0) = 0^2 + 1 = 1$$

(b) We require y when $x = 4$. The second part of the definition must be used because x lies between 2 and 6. Therefore

$$y(4) = 3(4) = 12$$

(c) We require y when $x = 2$. The value $x = 2$ occurs in the first part of the definition. Therefore

$$y(2) = 2^2 + 1 = 5$$

(d) We require y when $x = 7$. The final part of the function must be used. Therefore

$$y(7) = 2(7) + 1 = 15$$

Self-assessment questions 11.2

1. Explain what is meant by a function.
2. Explain the meaning of the terms 'dependent variable' and 'independent variable'.
3. Given $f(x)$, is the statement '$f(1/x)$ means $1/f(x)$' true or false?
4. Give an example of a function $f(x)$ such that $f(2) = f(3)$, that is the outputs for the inputs 2 and 3 are identical.

Exercise 11.2

1. Describe in words each of the following functions:

   (a) $h(t) = 10t$ (b) $g(x) = -x + 2$

   (c) $h(t) = 3t^4$ (d) $f(x) = \dfrac{4}{x^2}$

   (e) $f(x) = 3x^2 - 2x + 9$ (f) $f(x) = 5$
   (g) $f(x) = 0$

2. Describe in words each of the following functions:

   (a) $f(t) = 3t^2 + 2t$ (b) $g(x) = 3x^2 + 2x$

   Comment upon your answers.

3. Write the following functions using mathematical notation.

   (a) The input is cubed and the result is divided by 12.

   (b) The input is added to 3 and the result is squared.

   (c) The input is squared and added to 4 times the input. Finally, 10 is subtracted from the result.

   (d) The input is squared and added to 5. Then the input is divided by this result.

   (e) The input is cubed and then 1 is subtracted from the result.

   (f) 1 is subtracted from the input and the result is squared.

   (g) Twice the input is subtracted from 7 and the result is divided by 4.

   (h) The output is always -13 whatever the value of the input.

4. Given the function $A(n) = n^2 - n + 1$ evaluate

   (a) $A(2)$ (b) $A(3)$ (c) $A(0)$ (d) $A(-1)$

5. Given $y(x) = (2x - 1)^2$ evaluate

   (a) $y(1)$ (b) $y(-1)$ (c) $y(-3)$ (d) $y(0.5)$
   (e) $y(-0.5)$

6. The function f is given by $f(t) = 4t + 6$. Write expressions for

   (a) $f(t + 1)$ (b) $f(t + 2)$
   (c) $f(t + 1) - f(t)$ (d) $f(t + 2) - f(t)$

7. The function $f(x)$ is defined by $f(x) = 2x^2 - 3$. Write expressions for

   (a) $f(n)$ (b) $f(z)$ (c) $f(t)$ (d) $f(2t)$

   (e) $f\left(\dfrac{1}{z}\right)$ (f) $f\left(\dfrac{3}{n}\right)$ (g) $f(-x)$

   (h) $f(-4x)$ (i) $f(x + 1)$ (j) $f(2x - 1)$

8. Given the function $a(p) = p^2 + 3p + 1$ write an expression for $a(p+1)$. Verify that $a(p+1) - a(p) = 2p + 4$.

9. Sometimes the output from one function forms the input to another function. Suppose we have two functions f given by $f(t) = 2t$, and h given by $h(t) = t + 1$. $f(h(t))$ means that t is input to h, and the output from h is input to f. Evaluate

   (a) $f(3)$ (b) $h(2)$ (c) $f(h(2))$ (d) $h(f(3))$

10. The functions f and h are defined as in Question 9. Write down expressions for

    (a) $f(h(t))$ (b) $h(f(t))$

11. A function is defined by

$$f(x) = \begin{cases} x & 0 \le x < 1 \\ 2 & x = 1 \\ 1 & x > 1 \end{cases}$$

Evaluate

   (a) $f(0.5)$ (b) $f(1.1)$ (c) $f(1)$

11.3   The inverse of a function

We have described a function f as a rule which receives an input, say x, and generates an output, say y. We now consider the reversal of that process, namely finding a function which receives y as input and generates x as the output. If such a function exists it is called the inverse function of f. Figure 11.3 illustrates this schematically.

Figure 11.3
The inverse of f reverses the effect of f

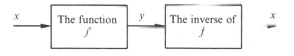

Worked examples

11.11   The functions f and g are defined by

$$f(x) = 2x \qquad g(x) = \frac{x}{2}$$

   (a) Verify that f is the inverse of g.
   (b) Verify that g is the inverse of f.

Solution   (a) The function g receives an input of x and generates an output of $x/2$, that is, it halves the input. In order to reverse the process, the inverse of g should receive $x/2$ as input and generate x as output. Now consider the function $f(x) = 2x$. This function doubles the input. Hence

$$f\left(\frac{x}{2}\right) = 2\left(\frac{x}{2}\right) = x$$

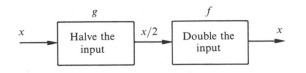

Figure 11.4
The function *f* is the
inverse of *g*

The function *f* has received $x/2$ as input and generated x as output. Hence *f* is the inverse of *g*. This is shown schematically in Figure 11.4.

(b) The function *f* receives x as input and generates $2x$ as output. In order to reverse the process, the inverse of *f* should receive $2x$ as input and generate x as output. Now $g(x) = x/2$, that is, the input is halved, and so

$$g(2x) = \frac{2x}{2} = x$$

Hence *g* is the inverse of *f*. This is shown schematically in Figure 11.5.

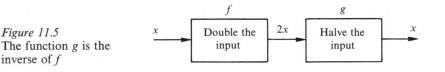

Figure 11.5
The function *g* is the
inverse of *f*

11.12   Find the inverse of the function $f(x) = 3x - 4$.

Solution   The function *f* multiplies the input by 3 and subtracts 4 from the result. To reverse the process, the inverse function, *g* say, must add 4 to the input and then divide the result by 3. Hence

$$g(x) = \frac{x + 4}{3}$$

11.13   Find the inverse of $h(t) = -\frac{1}{2}t + 5$.

Solution   The function *h* multiplies the input by $-\frac{1}{2}$ and then adds 5 to the result. Therefore the inverse function, *g* say, must subtract 5 from the input and then divide the result by $-\frac{1}{2}$. Hence

$$g(t) = \frac{(t - 5)}{-1/2} = -2(t - 5) = -2t + 10$$

There is an algebraic method of finding an inverse function which is often easier to apply. Suppose we wish to find the inverse of the function

$f(x) = 6 - 2x$. We let

$$y = 6 - 2x$$

and then transpose this for x. This gives

$$x = \frac{6 - y}{2}$$

Finally, we interchange x and y to give $y = (6 - x)/2$. This is the required inverse function. To summarize these stages:

KEY POINT

> To find the inverse of $y = f(x)$,
>
> - transpose the formula to make x the subject
> - interchange x and y
>
> The result is the required inverse function.

We shall meet some functions which do not have an inverse function. For example, consider the function $f(x) = x^2$. If 3 is the input, the output is 9. Now if -3 is the input, the output will also be 9 since $(-3)^2 = 9$. In order to reverse this process an inverse function would have to take an input of 9 and produce outputs of both 3 and -3. However, this contradicts the definition of a function which states that a function must have only *one* output for a given input. We say that $f(x) = x^2$ does not have an inverse function.

Self-assessment questions 11.3

1. Explain what is meant by the inverse of a function.
2. Explain why the function $f(x) = 4x^4$ does not possess an inverse function.

Exercise 11.3

1. Find the inverse of each of the following functions:

   (a) $f(x) = 3x$ (b) $f(x) = \dfrac{x}{4}$

   (c) $f(x) = x + 1$ (d) $f(x) = x - 3$
   (e) $f(x) = 3 - x$ (f) $f(x) = 2x + 6$

   (g) $f(x) = 7 - 3x$ (h) $f(x) = \dfrac{1}{x}$

   (i) $f(x) = \dfrac{3}{x}$ (j) $f(x) = -\dfrac{3}{4x}$

Test and assignment exercises 11

1. Given $r(t) = t^2 - t/2 + 4$ evaluate

   (a) $r(0)$ (b) $r(-1)$ (c) $r(2)$ (d) $r(3.6)$
   (e) $r(-4.6)$

2. A function is defined as $h(t) = t^2 - 7$:
   (a) state the dependent variable;
   (b) state the independent variable.

3. Given the functions $a(x) = x^2 + 1$ and $b(x) = 2x + 1$ write expressions for

   (a) $a(\alpha)$ (b) $b(t)$ (c) $a(2x)$ (d) $b\left(\dfrac{x}{3}\right)$

   (e) $a(x+1)$ (f) $b(x+h)$ (g) $b(x-h)$
   (h) $a(b(x))$ (i) $b(a(x))$

4. Find the inverse of each of the following functions:

   (a) $f(x) = \pi - x$ (b) $h(t) = \dfrac{t}{3} + 2$

   (c) $r(n) = \dfrac{1}{n}$ (d) $r(n) = \dfrac{1}{n-1}$

   (e) $r(n) = \dfrac{2}{n-1}$ (f) $r(n) = \dfrac{a}{n-b}$

   where a and b are constants.

5. Given $A(n) = n^2 + n - 6$ find expressions for

   (a) $A(n+1)$ (b) $A(n-1)$
   (c) $2A(n+1) - A(n) + A(n-1)$

12 Graphs of functions

Objectives

This chapter

- shows how to draw a graph of a function
- explains what is meant by the domain and range of a function
- explains how to use the graph of a function to solve equations

In Chapter 11 we introduced functions and represented them by algebraic equations. Another useful way of representing functions is pictorially, by means of **graphs**.

12.1 The x–y plane

We introduce horizontal and vertical axes as shown in Figure 12.1. These axes intersect at a point O called the **origin**. The horizontal axis is used to represent the independent variable, commonly x, and the vertical axis is used to represent the dependent variable, commonly y. The region shown is then referred to as the x–y plane.

A **scale** is drawn on both axes in such a way that at the origin $x = 0$ and $y = 0$. Positive x values lie to the right and negative x values to the left. Positive y values are above the origin, negative y values are below the origin. It is not essential that the scales on both axes are the same. Note that anywhere on the x axis, the value of y must be zero. Anywhere on the y axis the value of x must be zero. Each point in the plane corresponds to a specific value of x and y. We usually call the x value the x **coordinate** and call the y value the y **coordinate**. To refer to a specific point we give both its coordinates in brackets in the form (x, y), always giving the x coordinate first.

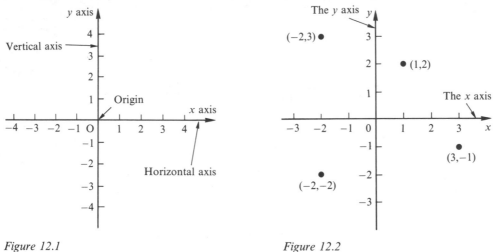

Figure 12.1
The x–y plane

Figure 12.2
Several points in the x–y plane

Worked examples

12.1   Draw the x–y plane and on it mark the points whose coordinates are
$(1, 2), (3, -1), (-2, 3), (-2, -2)$.

Solution   The plane is drawn in Figure 12.2, and the given points are indicated.
Note that the first coordinate in each bracket is the x coordinate.

Self-assessment questions 12.1

1. Explain the terms 'horizontal axis', 'vertical axis' and 'origin'.
2. When giving the coordinates of a point in the x–y plane, which coordinate is always given first?
3. State the coordinates of the origin.

Exercise 12.1

1. Plot and label the following points in the x–y plane:
$(-2, 0), (2, 0), (0, -2), (0, 2), (0, 0)$.

2. A set of points is plotted and then these points are joined by a straight line which is parallel to the x axis. What can you deduce about the coordinates of these points?

3. A set of points is plotted and then these points are joined by a straight line which is parallel to the y axis. What can you deduce about the coordinates of these points?

12.2   Plotting the graph of a function

The method of plotting a graph is best illustrated by example.

Worked examples

12.2   Plot a graph of $y = 2x - 1$ for $-3 \leq x \leq 3$.

Solution   We first calculate the value of y for several values of x. A table of x values and corresponding y values is drawn up as shown in Table 12.1.

The independent variable x varies from -3 to 3; the dependent variable y varies from -7 to 5. In order to accommodate all these values the scale on the x axis must vary from -3 to 3, and that on the y axis from -7 to 5. Each pair of values of x and y is represented by a unique point in the $x-y$ plane, with coordinates (x, y). Each pair of values in the table is plotted as a point and then the points are joined

Table 12.1
Values of x and y when
$y = 2x - 1$

x	-3	-2	-1	0	1	2	3
y	-7	-5	-3	-1	1	3	5

Figure 12.3
A graph of $y = 2x - 1$

to form the graph (Figure 12.3). By joining the points we see that, in this example, the graph is a straight line. We were asked to plot the graph for $-3 \le x \le 3$. Therefore, we have indicated each end point of the graph by a • to show that it is included. This follows the convention used for labelling closed intervals given in Chapter 1.

In the previous example, although we have used x and y as independent and dependent variables, clearly other letters could be used. The points to remember are that the axis for the independent variable is horizontal and the axis for the dependent variable is vertical. When coordinates are given the value of the independent variable is always given first.

Worked example

12.3 (a) Plot a graph of $y = x^2 - 2x + 2$ for $-2 \le x \le 3$.
   (b) Use your graph to determine which of the following points lie on the graph: $(1, 2)$, $(0, 2)$, and $(1, 1)$.

Solution (a) Table 12.2 gives values of x and the corresponding values of y. The independent variable is x and so the x axis is horizontal. The calculated points are plotted and joined. Figure 12.4 shows the graph. In this example the graph is not a straight line but rather a curve.
   (b) The point $(1, 2)$ has an x coordinate of 1 and a y coordinate of 2. The point is labelled by A on Figure 12.4. Points $(0, 2)$ and $(1, 1)$ are plotted and labelled by B and C. From the figure we can see that $(0, 2)$ and $(1, 1)$ lie on the graph.

x	-2	-1	0	1	2	3
y	10	5	2	1	2	5

Table 12.2
Values of x and y when
$y = x^2 - 2x + 2$

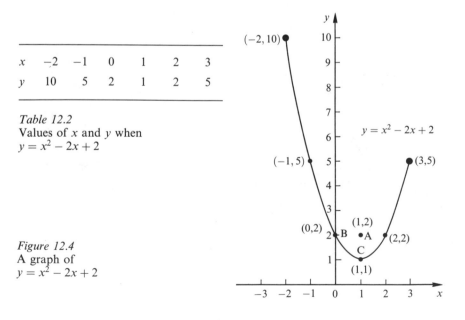

Figure 12.4
A graph of
$y = x^2 - 2x + 2$

Self-assessment questions 12.2

1. Which variable is plotted vertically on a graph - the independent or dependent?
2. Suppose we wished to plot a graph of the function $g = 4t^2 + 3t - 2$. Upon which axis would the variable t be plotted?

Exercise 12.2

1. Plot graphs of the following functions:

   (a) $y = f(x) = 2x + 7$ for $-4 \leq x \leq 4$.
   (b) $y = f(x) = -2x$ for values of x between -3 and 3 inclusive.
   (c) $y = f(x) = x$ for $-5 \leq x \leq 5$.
   (d) $y = f(x) = -x$ for $-5 \leq x \leq 5$.

2. Plot a graph of the function
   $y = f(x) = 7$, for $-2 \leq x < 2$.

3. A function is given by $y = f(x) = x^3 + x$ for $-3 \leq x \leq 3$.

   (a) Plot a graph of the function.
   (b) Determine which of the following points lie on the curve: $(0,1)$, $(1,0)$, $(1,2)$, and $(-1,-2)$

4. Plot, on the same axes, graphs of the functions $f(x) = x + 2$ for $-3 \leq x \leq 3$, and $g(x) = -\frac{1}{2}x + \frac{1}{2}$ for $-3 \leq x \leq 3$. State the coordinates of the point where the graphs intersect.

5. Plot a graph of the function $f(x) = x^2 - x - 1$ for $-2 \leq x \leq 3$. State the coordinates of the points where the curve cuts (a) the horizontal axis, (b) the vertical axis.

6. (a) Plot a graph of the function $y = 1/x$ for $0 < x \leq 3$.
   (b) Deduce the graph of the function $y = -1/x$ for $0 < x \leq 3$ from your graph in part (a).

7. Plot a graph of the function defined by

   $$r(t) = \begin{cases} t^2 & 0 \leq t < 3 \\ -2t + 15 & 3 < t < 7.5 \end{cases}$$

8. On the same axes plot a graph of $y = 2x$, $y = 2x + 1$ and $y = 2x + 3$, for $-3 \leq x \leq 3$. What observations can you make?

12.3   The domain and range of a function

The set of values which we allow the independent variable to take is called the **domain** of the function. If the domain is not actually specified in any particular example, it is taken to be the largest set possible. The set of values taken by the output is called the **range** of the function.

Worked examples

12.4   The function f is given by $y = f(x) = 2x$, for $1 \leq x \leq 3$.
(a) State the independent variable.
(b) State the dependent variable.
(c) State the domain of the function.
(d) Plot a graph of the function.
(e) State the range of the function.

Solution   (a) The independent variable is x.
(b) The dependent variable is y.
(c) Since we are given $1 \leq x \leq 3$ the domain of the function is the interval $[1, 3]$, that is, all values from 1 to 3 inclusive.
(d) The graph of $y = f(x) = 2x$ is shown in Figure 12.5.
(e) The range is the set of values taken by the output, y. From the graph we see that as x varies from 1 to 3 then y varies from 2 to 6. Hence the range of the function is $[2,6]$.

Figure 12.5 Graph of $y = f(x) = 2x$

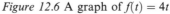

Figure 12.6 A graph of $f(t) = 4t$

12.5   Consider the function $y = f(t) = 4t$.

(a) State which is the independent variable and which is the dependent variable.
(b) Plot a graph of the function and give its domain and range.

Solution   (a) The independent variable is t and the dependent variable is y.
(b) No domain is specified so it is taken to be the largest set possible. The domain is $\mathbb{R}$, the set of all (real) numbers. However, it would be impractical to draw a graph whose domain was the whole extent of the t axis and so a selected portion is shown in Figure 12.6.

Similarly only a restricted portion of the range can be drawn although in this example the range is also $\mathbb{R}$.

12.6 Consider the function $y = f(x) = x^2 - 1$: (a) state the independent variable; (b) state the dependent variable; (c) state the domain; (d) plot a graph of f and determine the range.

Solution (a) The independent variable is x.
(b) The dependent variable is y.
(c) No domain is specified so it is taken to be the largest set possible. The domain is $\mathbb{R}$.
(d) A table of values and graph of $f(x) = x^2 - 1$ is shown in Figure 12.7. From the graph we see that the smallest value of y is -1 which occurs when $x = 0$. Whether x increases or decreases from 0, the value of y increases. The range is thus all values greater than or equal to -1. We can write the range using the set notation introduced in Chapter 1 as

$$\text{range} = \{y : y \in \mathbb{R}, y \geq -1\}$$

12.7 Given the function $y - 1/x$, state the domain and range.

Solution The domain is the largest set possible. All values of x are permissible except $x = 0$ since $1/0$ is not defined. It is never possible to divide by 0. Thus the domain comprises all real numbers except 0. A table of values and graph is shown in Figure 12.8. As x varies, y can take on any value except 0. In this example we see that the range is thus the same as the domain. We note that the graph is split at $x = 0$. For small, positive values of x, y is large and positive. The graph approaches the y axis for very small positive values of x. For small,

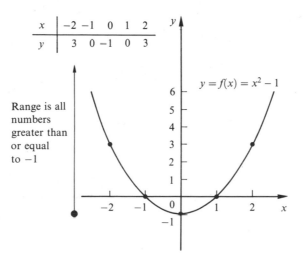

Figure 12.7
A graph of $f(x) = x^2 - 1$

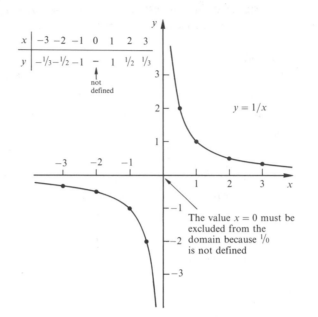

Figure 12.8
A graph of $y = 1/x$

negative values of x, y is large and negative. Again, the graph approaches the y axis for very small negative values of x. We say that the y axis is an **asymptote**. In general, if the graph of a function approaches a straight line we call that line an asymptote.

Self-assessment questions 12.3

1. Explain the terms 'domain of a function' and 'range of a function'.
2. Explain why the value $x = 3$ must be excluded from any domain of the function $f(x) = 7/(x - 3)$.
3. Give an example of a function for which we must exclude the value $x = -2$ from the domain.

Exercise 12.3

1. Plot graphs of the following functions. In each case state (i) the independent variable, (ii) the dependent variable, (iii) the domain, (iv) the range.

   (a) $y = x - 1$, $6 \leq x \leq 10$
   (b) $h = t^2 + 3$, $4 \leq t < 5$
   (c) $m = 3n - 2$, $-1 < n < 1$
   (d) $y = 2x + 4$
   (e) $X = y^3$
   (f) $k = 4r^2 + 5$, $5 \leq r \leq 10$

2. A function is defined by

   $$q(t) = \begin{cases} t^2 & 0 < t < 3 \\ 6 + t & 3 \leq t < 5 \\ -2t + 21 & 5 \leq t \leq 10.5 \end{cases}$$

   (a) Plot a graph of this function.
   (b) State the domain of the function.
   (c) State the range of the function.
   (d) Evaluate $q(1)$, $q(3)$, $q(5)$ and $q(7)$.

12.4   Solving equations using graphs

In Chapter 10 we used algebraic methods to solve equations. However, many equations cannot be solved exactly using such methods. In such cases it is often useful to use a graphical approach. The following examples illustrate the method of using graphs to solve equations.

Worked examples

12.8   Find graphically all solutions in the interval $[-3, 3]$ of the equation $x^2 + x - 3 = 0$.

Solution   We draw a graph of $y(x) = x^2 + x - 3$. Table 12.3 gives x and y values and Figure 12.9 shows a graph of the function. We have plotted $y = x^2 + x - 3$ and wish to solve $0 = x^2 + x - 3$. Thus we read from the graph the coordinates of the points at which $y = 0$. Such points must be on the x axis. From the graph the points are $(1.34, 0)$ and $(-2.30, 0)$. So the solutions of $x^2 + x - 3 = 0$ are $x = 1.34$ and $x = -2.30$. Note that these are approximate solutions, being dependent upon the accuracy of the graph and the x and y scales used. Increased accuracy can be achieved by drawing an enlargement of the graph

Table 12.3
Values of x and y when $y = x^2 + x - 3$

x	-3	-2	-1	0	1	2	3
y	3	-1	3	3	-1	3	9

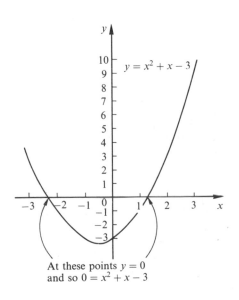

Figure 12.9
A graph of $y = x^2 + x - 3$

At these points $y = 0$ and so $0 = x^2 + x - 3$

around the values $x = 1.34$ and $x = -2.30$. So, for example, we could draw $y = x^2 + x - 3$ for $1.2 \le x \le 1.5$ and $y = x^2 + x - 3$ for $-2.4 \le x \le -2.2$.

12.9 Find solutions in the interval $[-2,2]$ of the equation $x^3 = 2x + \frac{1}{2}$ using a graphical method.

Solution The problem of solving $x^3 = 2x + \frac{1}{2}$ is identical to that of solving $x^3 - 2x - \frac{1}{2} = 0$. So we plot a graph of $y = x^3 - 2x - \frac{1}{2}$ for $-2 \le x \le 2$ and then locate the points where the curve cuts the x axis, that is, where the y coordinate is 0. Table 12.4 gives x and y values and Figure 12.10 shows a graph of the function. We now consider points on the graph where the y coordinate is zero. These points are marked A, B and C. Their x coordinates are -1.27, -0.26 and 1.53. Hence the solutions of $x^3 = 2x + \frac{1}{2}$ are approximately $x = -1.27$, $x = -0.26$ and $x = 1.53$.

Table 12.4
Values of x and y when $y = x^3 - 2x - \frac{1}{2}$

x	-2	-1.5	-1	-0.5	0	0.5	1	1.5	2
y	-4.5	-0.875	0.5	0.375	-0.5	-1.375	-1.5	-0.125	3.5

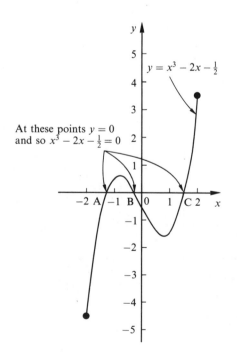

Figure 12.10
A graph of
$y = x^3 - 2x - \frac{1}{2}$

12.10 (a) On the same axes draw graphs of $y = x^3$ and $y = 2x + \frac{1}{2}$ for $-2 \le x \le 2$.

(b) Note the x coordinates of the points where the two graphs intersect. By referring to Worked Example 12.9 what do you conclude? Can you explain your findings?

Solution (a) Table 12.5 gives x values and the corresponding values of x^3 and $2x + \frac{1}{2}$. Figure 12.11 shows a graph of $y = x^3$ together with a graph of $y = 2x + \frac{1}{2}$.

Table 12.5
Values of x and y when
$y = x^3$ and $y = 2x + \frac{1}{2}$

x	-2	-1.5	-1	-0.5	0	0.5	1	1.5	2
x^3	-8	-3.375	-1	-0.125	0	0.125	1	3.375	8
$2x + \frac{1}{2}$	-3.5	-2.5	-1.5	-0.5	0.5	1.5	2.5	3.5	4.5

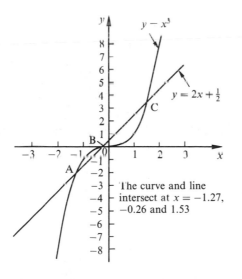

The curve and line intersect at $x = -1.27$, -0.26 and 1.53

Figure 12.11
Graphs of $y = x^3$ and
$y = 2x + \frac{1}{2}$

(b) The two graphs intersect at A, B and C. The x coordinates of these points are -1.27, -0.26 and 1.53. We note from Worked Example 12.9 that these values are the solutions of $x^3 = 2x + \frac{1}{2}$. We explain this as follows. Where $y = x^3$ and $y = 2x + \frac{1}{2}$ intersect, their y values are identical and so at these points

$$x^3 = 2x + \frac{1}{2}$$

By reading the x coordinates at these points of intersection we are finding those x values for which $x^3 = 2x + \frac{1}{2}$.

Self-assessment questions 12.4

1. Explain how the graph of the function f can be used to solve the equation (a) $f(x) = 0$, (b) $f(x) = 1$.

Exercise 12.4

1. Plot a graph of

$$y = x^2 + \frac{3x}{2} - 2$$

for $-3 \le x \le 2$. Hence solve the equation

$$x^2 + \frac{3x}{2} - 2 = 0$$

2. (a) On the same axes plot the graphs of $y = x^2$ and $y = 2x + 1$ for $-3 \le x \le 3$.

   (b) Use your graphs to solve the equation $x^2 = 2x + 1$.

   (c) By drawing an additional line on the graph find solutions of $x^2 + x - 4 = 0$.

Test and assignment exercises 12

1. Plot a graph of each of the following functions. In each case state the independent variable, the dependent variable, the domain and the range.

   (a) $y = f(x) = 3x - 7$
   (b) $y = f(x) = 3x - 7$, $-2 \le x \le 2$
   (c) $y = 4x^2$, $x \ge 0$
   (d) $y = \dfrac{1}{x^2}$, $x \ne 0$

2. A function is defined by

   $$f(x) = \begin{cases} 2x + 4 & 0 \le x \le 4 \\ 6 & 4 < x < 6 \\ x & 6 \le x \le 8 \end{cases}$$

   (a) Plot a graph of this function, stating its domain and range.
   (b) Evaluate $f(0)$, $f(4)$ and $f(6)$.

3. Plot graphs of the function $f(x) = x + 4$ and $g(x) = 2 - x$ for values of x between -4 and 4. Hence solve the simultaneous equations $y - x = 4$, $y + x = 2$.

4. Plot a graph of the function $f(x) = 3x^2 + 10x - 8$ for values of x between -5 and 3. Hence solve the equation $3x^2 + 10x - 8 = 0$.

5. Plot a graph of the function $f(t) = t^3 + 2t^2 - t - 2$ for $-3 \le t \le 3$. Hence solve the equation $t^3 + 2t^2 - t - 2 = 0$.

6. Plot a graph of $h = 2 + 2x - x^2$ for $-3 \le x \le 3$. Hence solve the equation $2 + 2x - x^2 = 0$.

7. (a) On the same axes plot graphs of $y = x^3$ and $y = 5x^2 - 3$ for $-2 \le x \le 2$.
   (b) Use your graphs to solve the equation $x^3 - 5x^2 + 3 = 0$.

8. (a) Plot a graph of $y = x^3 - 2x^2 - x + 1$ for $-2 \le x \le 3$.
   (b) Hence find solutions of $x^3 - 2x^2 - x + 1 = 0$.
   (c) Use your graph to solve $x^3 - 2x^2 - x + 2 = 0$.

13 The straight line

Objectives	This chapter
	• describes some special properties of straight line graphs
	• explains the equation $y = mx + c$
	• explains the terms 'vertical intercept' and 'gradient'

13.1   Straight line graphs

In the previous chapter we explained how to draw graphs of functions. Several of the resulting graphs were straight lines. In this chapter we focus specifically on the straight line and some of its properties.

Any equation of the form $y = mx + c$, where m and c are constants, will have a straight line graph. For example,

$$y = 3x + 7 \qquad y = -2x + \frac{1}{2} \qquad y = 3x - 1.5$$

will all result in straight line graphs. It is important to note that the variable x only occurs to the power 1. The values of m and c may be zero, so that $y = -3x$ is a straight line for which the value of c is zero, and $y = 17$ is a straight line for which the value of m is zero.

KEY POINT

> Any straight line has an equation of the form $y = mx + c$ where m and c are constants.

Worked examples

13.1 Which of the following equations have straight line graphs? For those that do, identify the values of m and c.

$$\text{(a) } y = 7x + 5 \text{ (b) } y = \frac{13x - 5}{2} \text{ (c) } y = 3x^2 + x \text{ (d) } y = -19$$

Solution   (a) $y = 7x + 5$ is an equation of the form $y = mx + c$ where $m = 7$ and $c = 5$. This equation has a straight line graph.

(b) The equation

$$y = \frac{13x - 5}{2}$$

can be written as $y = \frac{13}{2}x - \frac{5}{2}$ which is in the form $y = mx + c$ with $m = \frac{13}{2}$ and $c = -\frac{5}{2}$. This has a straight line graph.

(c) $y = 3x^2 + x$ contains the term x^2. Such a term is not allowed in the equation of a straight line. The graph will not be a straight line.

(d) $y = -19$ is in the form of the equation of a straight line for which $m = 0$ and $c = -19$.

13.2   Plot each of the following graphs: $y = 2x + 3$, $y = 2x + 1$, and $y = 2x - 2$. Comment upon the resulting lines.

Solution   The three graphs are shown in Figure 13.1. Note that all three graphs have the same steepness or **slope**. However, each one cuts the vertical axis at a different point. This point can be obtained directly from the equation $y = 2x + c$ by looking at the value of c. For example,

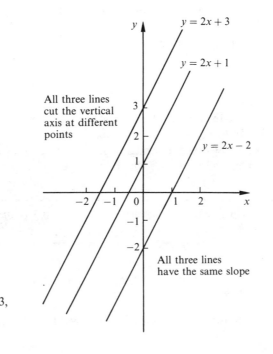

Figure 13.1
Graphs of $y = 2x + 3$,
$y = 2x + 1$ and
$y = 2x - 2$

$y = 2x + 1$ intersects the vertical axis at $y = 1$. The graph of $y = 2x + 3$ cuts the vertical axis at $y = 3$, and the graph of $y = 2x - 2$ cuts this axis at $y = -2$.

The point where a graph cuts the vertical axis is called the **vertical intercept**. The vertical intercept can be obtained from $y = mx + c$ by looking at the value of c.

KEY POINT

In the equation $y = mx + c$ the value of c gives the y coordinate of the point where the line cuts the vertical axis.

Worked example

13.3   On the same graph plot the following. $y = x + 2$, $y = 2x + 2$, $y = 3x + 2$. Comment upon the graphs.

Solution   The three straight lines are shown in Figure 13.2. Note that the steepness of the line is determined by the coefficient of x, with $y = 3x + 2$ being steeper than $y = 2x + 2$, and this in turn being steeper than $y = x + 2$.

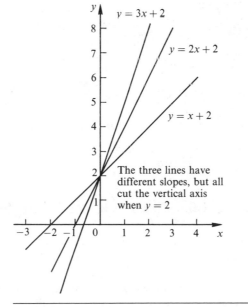

Figure 13.2
Graphs of $y = x + 2$,
$y = 2x + 2$, and
$y = 3x + 2$

The value of m in the equation $y = mx + c$ determines the steepness of the straight line. The larger the value of m, the steeper is the line. The value m is known as the slope or **gradient** of the straight line.

KEY POINT
> In the equation $y = mx + c$ the value m is known as the gradient and is a measure of the steepness of the line.

If m is positive, the line will rise as we move from left to right. If m is negative, the line will fall. If $m = 0$ the line will be horizontal. We say it has zero gradient. These points are summarized in Figure 13.3.

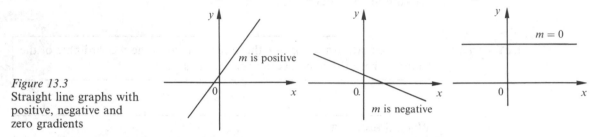

Figure 13.3
Straight line graphs with positive, negative and zero gradients

---

Self-assessment questions 13.1

1. Give the standard form of the equation of a straight line.
2. Explain the meaning of m and c in the equation $y = mx + c$.

Exercise 13.1

1. Identify, without plotting, which of the following functions will give straight line graphs:

   (a) $y = 3x + 9$ (b) $y = -9x + 2$
   (c) $y = -6x$ (d) $y = x^2 + 3x$
   (e) $y = 17$ (f) $y = x^{-1} + 7$

2. Identify the gradient and vertical intercept of each of the following lines:

   (a) $y = 9x - 11$ (b) $y = 8x + 1.4$

   (c) $y = \frac{1}{2}x - 11$ (d) $y = 17 - 2x$

   (e) $y = \frac{2x + 1}{3}$ (f) $y = \frac{4 - 2x}{5}$

   (g) $y = 3(x - 1)$ (h) $y = 4$

13.2   Finding the equation of a straight line from its graph

If we are given the graph of a straight line it is often necessary to find its equation, $y = mx + c$. This amounts to finding the values of m and c. Finding the vertical intercept is straightforward because we can look directly for the point where the line cuts the y axis. The y coordinate of this point gives the value of c. The gradient m can be determined from knowledge of any two points on the line using the formula

$$\text{gradient} = \frac{\text{difference between their } y \text{ coordinates}}{\text{difference between their } x \text{ coordinates}}$$

Worked examples

13.4   A straight line graph is shown in Figure 13.4. Determine its equation.

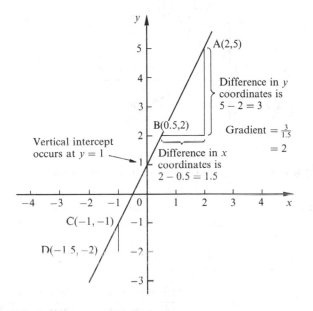

Figure 13.4
Graph for Example 13.4

Solution   We require the equation of the line in the form $y = mx + c$. From the graph it is easy to see that the vertical intercept occurs at $y = 1$. Therefore the value of c is 1. To find the gradient m we choose any two points on the line. We have chosen the point A with coordinates $(2,5)$ and the point B with coordinates $(0.5, 2)$. The difference between their y coordinates is then $5 - 2 = 3$. The difference between their x coordinates is $2 - 0.5 = 1.5$. Then

$$\text{gradient} = \frac{\text{difference between their } y \text{ coordinates}}{\text{difference between their } x \text{ coordinates}}$$

$$= \frac{3}{1.5} = 2$$

The gradient m is equal to 2. Note that as we move from left to right the line is rising and so the value of m is positive. The equation of the line is then $y = 2x + 1$. There is nothing special about the points A and B. Any two points are sufficient to find m. For example, using the points C with coordinates $(-1, -1)$ and D with coordinates $(-1.5, -2)$ we would find

$$\text{gradient} = \frac{\text{difference between their } y \text{ coordinates}}{\text{difference between their } x \text{ coordinates}}$$

$$= \frac{-1 - (-2)}{-1 - (-1.5)}$$

$$= \frac{1}{0.5} = 2, \text{ as before.}$$

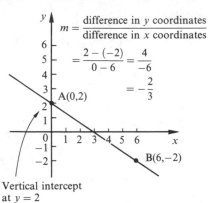

Figure 13.5
Graph for Example 13.5

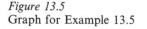

Vertical intercept
at $y = 2$

13.5   A straight line graph is shown in Figure 13.5. Find its equation.

Solution   We need to find the equation in the form $y = mx + c$. From the graph we see immediately that the value of c is 2. To find the gradient we have selected any two points, $A(0, 2)$ and $B(6, -2)$. The difference between their y coordinates is $2 - (-2) = 4$. The difference between their x coordinates is $0 - 6 = -6$. Then

$$\text{gradient} = \frac{\text{difference between their } y \text{ coordinates}}{\text{difference between their } x \text{ coordinates}}$$

$$= \frac{4}{-6}$$

$$= -\frac{2}{3}$$

The equation of the line is therefore $y = -\frac{2}{3}x + 2$. Note in particular that because the line is sloping downwards as we move from left to right, the gradient is negative. Note also that the coordinates of A and B both satisfy the equation of the line, that is, for $A(0, 2)$,

$$2 = -\frac{2}{3}(0) + 2$$

and for $B(6, -2)$,

$$-2 = -\frac{2}{3}(6) + 2$$

The coordinates of any other point on the line must also satisfy the equation.

The point noted at the end of example 13.5 is important:

KEY POINT   If the point (a, b) lies on the line $y = mx + c$ then this equation is satisfied by letting $x = a$ and $y = b$.

Worked example

13.6   Find the equation of the line shown in Figure 13.6.

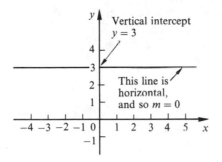

Figure 13.6
Graph for Example 13.6

Solution   We are required to express the equation in the form $y = mx + c$. From the graph we notice that the line is horizontal. This means that its gradient is 0, that is, $m = 0$. Furthermore the line cuts the vertical axis at $y = 3$ and so the equation of the line is $y = 0x + 3$ or simply $y = 3$.

It is not necessary to sketch a graph in order to find the equation. Consider the following example which illustrates an algebraic method.

Worked example

13.7   A straight line passes through $A(7, 1)$ and $B(-3, 2)$. Find its equation.

Solution   The equation must be of the form $y = mx + c$. The gradient of the line can be found from

$$\text{gradient} = m = \frac{\text{difference between their } y \text{ coordinates}}{\text{difference between their } x \text{ coordinates}}$$

$$= \frac{1 - 2}{7 - (-3)}$$

$$= \frac{-1}{10}$$

$$= -0.1$$

Hence $y = -0.1x + c$. We can find c by noting that the line passes through (7,1), that is, the point where $x = 7$ and $y = 1$. Substituting these values into the equation $y = -0.1x + c$ gives

$$1 = -0.1(7) + c$$

so that $c = 1 + 0.7 = 1.7$. Therefore the equation of the line is $y = -0.1x + 1.7$.

Self-assessment questions 13.2

1. State the formula for finding the gradient of a straight line when two points upon it are known. If the two points are (x_1, y_1) and (x_2, y_2) write down an expression for the gradient.
2. Explain how the value of c in the equation $y = mx + c$ can be found by inspecting the straight line graph.

Exercise 13.2

1. A straight line passes through the two points (1,7) and (2,9). Sketch a graph of the line and find its equation.

2. Find the equation of the line which passes through the two points (2,2) and (3,8).

3. Find the equation of the line which passes through (8,2) and (−2,2).

4. Find the equation of the straight line which has gradient 1 and passes through the origin.

5. Find the equation of the straight line which has gradient −1 and passes through the origin.

6. Find the equation of the straight line passing through (−1, 6) with gradient 2.

7. Which of the following points lie on the line $y = 4x - 3$?

   (a) (1,2) (b) (2,5) (c) (5, 17)
   (d) (−1,−7) (e) (0,2)

13.3   Gradients of curves

Figure 13.7 shows a graph of $y = x^2$. If you study the graph you will notice that as we move from left to right, at some points the y values are decreasing, whilst at others the y values are increasing. It is intuitively obvious that the slope of the curve changes from point to point. At some points, such as A, the curve appears quite steep and falling. At points such as B the curve appears quite steep and rising. Unlike a straight line, the slope of a curve is not fixed but changes as we move from one point to another. A useful way of measuring the slope at any point is to draw a **tangent** to the curve at that point. The tangent is a straight line which just touches the curve at the point of interest. In Figure 13.7 a tangent to the curve $y = x^2$ has been drawn at the point (2,4). If we calculate the gradient of this tangent, this gives the gradient of the curve at the point (2,4).

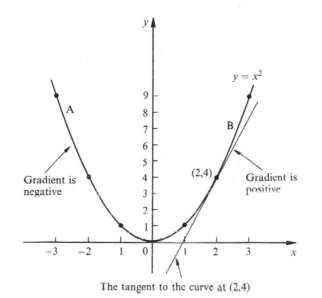

Figure 13.7
A graph of $y - x^2$

The tangent to the curve at (2,4)

KEY POINT

The gradient of a curve at any point is equal to the gradient of the tangent at that point.

Worked example

13.8   (a) Plot a graph of $y = x^2 - x$ for values of x between -2 and 4.

(b) Draw in tangents at the points $A(-1,2)$ and $B(3,6)$.

(c) By calculating the gradients of these tangents find the gradient of the curve at A and at B.

Solution   (a) A table of values and the graph is shown in Figure 13.8.

(b) We now draw tangents at A and B. At present, the best we can do is estimate these by eye.

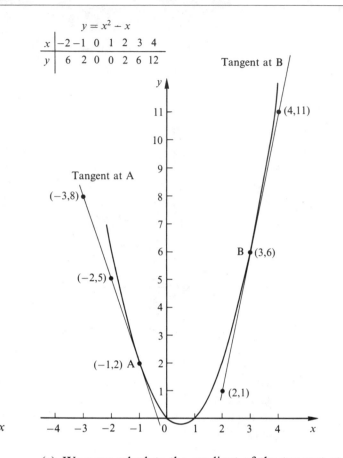

Figure 13.8
A graph of $y = x^2 - x$

(c) We now calculate the gradient of the tangent at A. We select any two points on the tangent and calculate the difference between their y coordinates and the difference between their x coordinates. We have chosen the points $(-3, 8)$ and $(-2, 5)$. Referring to Figure 13.8 we see that

$$\text{gradient of tangent at } A = \frac{8 - 5}{-3 - (-2)} = \frac{3}{-1}$$
$$= -3$$

Hence the gradient of the curve at A is -3. Similarly, to find the gradient of the tangent at B we have selected two points on this tangent, namely $(4, 11)$ and $(2, 1)$. We find

$$\text{gradient of tangent at } B = \frac{11 - 1}{4 - 2} = \frac{10}{2}$$
$$= 5$$

Hence the gradient of the tangent at B is 5.

Clearly, the accuracy of our answer depends to a great extent upon how well we can draw and measure the gradient of the tangent.

___

Worked example

13.9 (a) Sketch a graph of the curve $y = x^3$ for $-2 \leq x \leq 2$.

(b) Draw the tangent to the graph at the point where $x = 1$.

(c) Estimate the gradient of this tangent and find its equation.

Solution (a) A graph is shown in Figure 13.9.

(b) The tangent has been drawn at $x = 1$.

(c) Let us write the equation of the tangent as $y = mx + c$. Two points on the tangent have been selected in order to estimate the gradient. These are $(2, 4)$ and $(-1, -5)$. From these we find

$$\text{gradient of tangent is approximately } \frac{4 - (-5)}{2 - (-1)} = \frac{9}{3} = 3$$

Therefore $m = 3$. The value of c is found by noting that the vertical intercept of the tangent is -2. The equation of the tangent is then $y = 3x - 2$.

___

Of course, this method will usually result in an approximation based upon how well we have drawn the graph and its tangent. A much more precise method for calculating gradients of curves is given in Chapter 21.

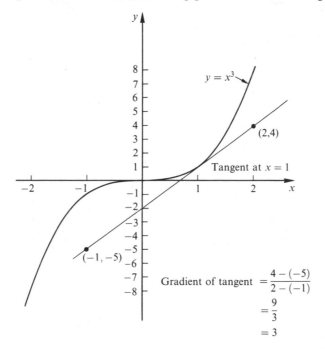

$y = x^3$

$(2, 4)$

Tangent at $x = 1$

$(-1, -5)$

$$\text{Gradient of tangent} = \frac{4 - (-5)}{2 - (-1)}$$
$$= \frac{9}{3}$$
$$= 3$$

Figure 13.9
Graph of $y = x^3$

Self-assessment questions 13.3

1. Explain what is meant by the 'tangent' to a curve at a point.
2. Explain how a tangent is used to determine the gradient of a curve.

Exercise 13.3

1. Draw the graph of $y = 2x^2 - 1$ for values of x between -3 and 3. By drawing a tangent, estimate the gradient of the curve at $A(2,7)$ and $B(-1,1)$.

2. Draw the graph of $y = -2x^2 + 2$ for values of x between -3 and 3. Draw the tangent at the point where $x = 1$ and calculate its equation.

Test and assignment exercises 13

1. Which of the following will have straight line graphs?

   (a) $y = 2x - 11$ (b) $y = 5x + 10$
   (c) $y = x^2 - 1$ (d) $y = -3 + 3x$
   (e) $y = \dfrac{2x + 3}{2}$

   For each straight line, identify the gradient and vertical intercept.

2. Find the equation of the straight line which passes through the points $(1,11)$ and $(2,18)$. Show that the line also passes through $(-1,-3)$.

3. Find the equation of the line which has gradient -2 and passes through the point $(1,1)$.

4. Find the equation of the line which passes through $(-1,5)$ and $(1,5)$. Does the line also pass through $(2,6)$?

5. Draw a graph of $y = -x^2 + 3x$ for values of x between -3 and 3. By drawing in tangents estimate the gradient of the curve at the points $(-2,-10)$ and $(1,2)$.

6. Find the equations of the lines passing through the origin with gradients (a) -2, (b) -4, (c) 4.

14 The exponential function

| Objectives |

This chapter

- shows how to simplify exponential expressions
- describes the form of the exponential function
- illustrates graphs of exponential functions
- lists the properties of the exponential function
- shows how to solve equations with exponential terms using a graphical technique

14.1 Exponential expressions

The letter e always denotes the constant 2.71828 ...

An expression of the form a^x is called an **exponential expression**. The number a is called the **base** and x is the **power** or **index**. For example, 4^x and 0.5^x are both exponential expressions. The first has base 4 and the second has base 0.5.

One of the most commonly used values for the base in an exponential expression is 2.71828.... This irrational number is denoted by the letter e. The expression e^x is found to occur in many natural phenomena, for example, population growth, spread of bacteria and radioactive decay. Scientific calculators are pre-programmed to evaluate e^x for any value of x. Usually the button is marked e^x. Use the next example to check you can use this facility on your calculator.

Worked example

14.1 Use a scientific calculator to evaluate

(a) e^2 (b) e^3 (c) $e^{1.3}$ (d) $e^{-1.3}$

Solution We use the e^x button giving:

(a) $e^2 = 7.3891$

(b) $e^3 = 20.0855$

(c) $e^{1.3} = 3.6693$

(d) $e^{-1.3} = 0.2725$

Exponential expressions can be simplified using the normal rules of algebra. The laws of indices apply to exponential expressions. We note

$$e^a e^b = e^{a+b}$$

$$\frac{e^a}{e^b} = e^{a-b}$$

$$e^0 = 1$$

$$(e^a)^b = e^{ab}$$

Worked examples

14.2   Simplify the following exponential expressions:

(a) $e^2 e^4$ (b) $\dfrac{e^4}{e^3}$ (c) $(e^2)^{2.5}$

Solution   (a) $e^2 e^4 = e^{2+4} = e^6$

(b) $\dfrac{e^4}{e^3} = e^{4-3} = e^1 = e$

(c) $(e^2)^{2.5} = e^{2 \times 2.5} = e^5$

14.3   Simplify the following exponential expressions:

(a) $e^x e^{3x}$ (b) $\dfrac{e^{4t}}{e^{3t}}$ (c) $(e^t)^4$

Solution   (a) $e^x e^{3x} = e^{x+3x} = e^{4x}$

(b) $\dfrac{e^{4t}}{e^{3t}} = e^{4t-3t} = e^t$

(c) $(e^t)^4 = e^{4t}$

14.4   Simplify

(a) $\sqrt{e^{6t}}$ (b) $e^{3y}(1 + e^y) - e^{4y}$ (c) $(2e^t)^2(3e^{-t})$

Solution   (a) $\sqrt{e^{6t}} = (e^{6t})^{1/2} = e^{6t/2} = e^{3t}$

(b) $e^{3y}(1 + e^y) - e^{4y} = e^{3y} + e^{3y}e^y - e^{4y}$

$$= e^{3y} + e^{4y} - e^{4y}$$

$$= e^{3y}$$

(c) $(2c')^2(3e^{-t}) = (2e')(2e')(3e^{-t})$

$\qquad = 2.2.3.e^t e^t e^{-t}$

$\qquad = 12e^{t+t-t}$

$\qquad = 12e^t$

14.5 (a) Verify that

$$(e^t + 1)^2 = e^{2t} + 2e^t + 1$$

(b) Hence simplify $\sqrt{e^{2t} + 2e^t + 1}$.

Solution (a) $(e^t + 1)^2 = (e^t + 1)(e^t + 1)$

$\qquad = e^t e^t + e^t + e^t + 1$

$\qquad = e^{2t} + 2e^t + 1$

(b) $\sqrt{e^{2t} + 2e^t + 1} = \sqrt{(e^t + 1)^2} = e^t + 1$

Exercise 14.1

1. Use a scientific calculator to evaluate

(a) $e^{2.3}$ (b) $e^{1.9}$ (c) $e^{-0.6}$ (d) $\dfrac{1}{e^{-2} + 1}$

(e) $\dfrac{3e^2}{e^2 - 10}$

2. Simplify

(a) $e^3.e^7$ (b) $\dfrac{e^7}{e^4}$ (c) $\dfrac{e^{-2}}{e^3}$ (d) $\dfrac{e^3}{e^{-1}}$

(e) $\dfrac{(4e)^2}{(2e)^3}$ (f) $e^2\left(e + \dfrac{1}{e}\right) - e$

(g) $e^{1.5}e^{2.7}$ (h) $\sqrt{e}\sqrt{4e}$

3. Simplify as far as possible

(a) $\dfrac{e^t}{e^3}$ (b) $\dfrac{2e^{3x}}{4e^x}$

(c) $\dfrac{(e^x)^3}{3e^x}$ (d) $e^x + 2e^{-x} + e^x$

(e) $\dfrac{e^x e^y e^z}{e^{x/2-y}}$ (f) $\dfrac{(e^{t/3})^6}{e^t + e^t}$

(g) $\dfrac{e^t}{e^{-t}}$ (h) $\dfrac{1 + e^{-t}}{e^{-t}}$

(i) $(e^z e^{-z/2})^2$ (j) $\dfrac{e^{-3+t}}{2e^t}$

(k) $\left(\dfrac{e^{-1}}{e^{-r}}\right)^{-1}$

4. The hyperbolic functions $\sinh x$ and $\cosh x$ are defined by

$$\sinh x = \dfrac{e^x - e^{-x}}{2} \qquad \cosh x = \dfrac{e^x + e^{-x}}{2}$$

Simplify

(a) $\sinh x + \cosh x$

(b) $\cosh x - \sinh x$

(c) $(\cosh x)^2 - (\sinh x)^2$

14.2   The exponential function and its graph

An exponential function has the form

$$y = a^x$$

where a is a positive constant. Note that the independent variable x is a power, or index, of a. The dependent variable is y. The function $y = e^x$ is found to occur in many natural phenomena and so is often referred to as the **exponential function**. Table 14.1 gives values of x and e^x. Using Table 14.1, a graph of $y = e^x$ can be drawn. This is shown in Figure 14.1. From the table and graph we note the following important properties of $y = e^x$.

(i) The exponential function is never negative.
(ii) When $x = 0$, the function value is 1.
(iii) As x increases, then e^x increases. This is known as **exponential growth**.
(iv) By choosing x large enough, the value of e^x can be made larger than any given number. We say e^x increases without bound as x increases.
(v) As x becomes large and negative, e^x gets nearer and nearer to 0.

Table 14.1

x	-3	-2	-1	0	1	2	3
e^x	0.0498	0.1353	0.3679	1	2.7183	7.3891	20.086

Figure 14.1
A graph of $y = e^x$

Worked example

14.6   Sketch a graph of $y = 2e^x$ and $y = e^{2x}$ for $-3 \le x \le 3$.

Solution   Table 14.2 gives values of $2e^x$ and e^{2x} and the graphs are shown in Figure 14.2.

Table 14.2

x	−3	−2	−1	0	1	2	3
$2e^x$	0.0996	0.2707	0.7358	2	5.4366	14.7781	40.1711
e^{2x}	0.0025	0.0183	0.1353	1	7.3891	54.5982	403.429

Figure 14.2
Graphs of $y = e^{2x}$ and $y = 2e^x$

We now turn attention to an associated function: $y = e^{-x}$. Values are listed in Table 14.3 and the function is illustrated in Figure 14.3. From the table and the graph we note the following properties of $y = e^{-x}$.

(i) The function is never negative.
(ii) When $x = 0$, the function has a value of 1.
(iii) As x increases, then e^{-x} decreases, getting nearer to 0. This is known as **exponential decay**.

Table 14.3

x	−3	−2	−1	0	1	2	3
$-x$	3	2	1	0	−1	−2	−3
e^{-x}	20.086	7.3891	2.7183	1	0.3679	0.1353	0.0498

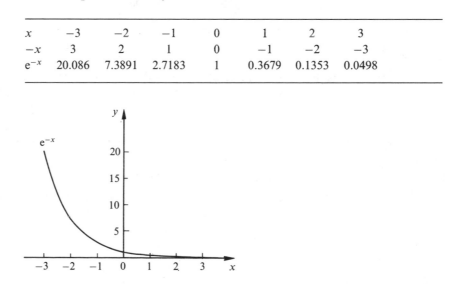

Figure 14.3
Graph of $y = e^{-x}$

Self-assessment questions 14.2

1. List properties which are common to $y = e^x$ and $y = e^{-x}$.
2. By choosing x large enough, then e^{-x} can be made to be negative. True or false?

Exercise 14.2

1. Draw up a table of values of $y = e^{3x}$ for x between -1 and 1 at intervals of 0.2. Sketch the graph of $y = e^{3x}$. Comment upon its shape and properties.

2. A species of animal has population $P(t)$ at time t given by

   $$P = 10 - 5e^{-t}$$

   (a) Sketch a graph of P against t for $0 \le t \le 5$.

   (b) What is the size of the population of the species as t becomes very large?

14.3 **Solving equations involving exponential terms using a graphical method**

Many equations involving exponential terms can be solved using graphs. A graphical method will yield an approximate solution which can then be refined by choosing a smaller interval for the domain. The following examples illustrate the technique.

Worked example

14.7 (a) Plot $y = e^{x/2}$ and $y = 2e^{-x}$ for $-1 \le x \le 1$.

(b) Hence solve the equation

$$e^{x/2} - 2e^{-x} = 0$$

Solution (a) A table of values is drawn up for $e^{x/2}$ and $2e^{-x}$ so that the graphs can be drawn. Table 14.4 lists the values and Figure 14.4 shows the graphs of the functions.

Table 14.4

x	-1	-0.75	-0.5	-0.25	0	0.25	0.5	0.75	1
$e^{x/2}$	0.607	0.687	0.779	0.882	1	1.133	1.284	1.455	1.649
$2e^{-x}$	5.437	4.234	3.297	2.568	2	1.558	1.213	0.945	0.736

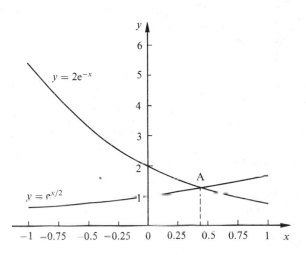

Figure 14.4
The graphs of $y = e^{x/2}$
and $y = 2e^{-x}$ intersect at
A

(b) We note that the graphs intersect; the point of intersection is A. The equation

$$e^{x/2} - 2e^{-x} = 0$$

is equivalent to

$$e^{x/2} = 2e^{-x}$$

The graphs of $y = e^{x/2}$ and $y = 2e^{-x}$ intersect at A, where $x = 0.44$. Hence $x = 0.44$ is an approximate solution of $e^{x/2} - 2e^{-x} = 0$. The exact answer can be shown to be 0.46 (2 d.p.).

Exercise 14.3

1. (a) Plot $y = 15 - x^2$ and $y = e^x$ for $0 \le x \le 3$.

   (b) Hence find an approximate solution to

   $$e^x + x^2 = 15$$

2. (a) Plot $y = 12x^2 - 1$ and $y = 3e^{-x}$ for $-1 \le x \le 1$.

   (b) Hence state two approximate solutions of $3e^{-x} = 12x^2 - 1$.

3. By drawing $y = e^x$ for $-3 \le x \le 3$ and additional appropriate lines solve the following equations:

   (a) $e^x + x = 0$

   (b) $e^x - 1.5 = 0$

   (c) $e^x - x - 5 = 0$

   (d) $\dfrac{e^x}{2} + \dfrac{x}{2} - 5 = 0$

Test and assignment exercises 14

1. Use a calculator to evaluate

   (a) $e^{-3.1}$ (b) $e^{0.2}$ (c) $\dfrac{1}{e}$

2. (a) Sketch the graphs of

   $y = e^{-x}$ and $y = \dfrac{x}{2}$

   for $0 \le x \le 2$.

   (b) Hence solve the equation

   $2e^{-x} - x = 0$

3. Simplify

   (a) $e^t e^{t/2}$ (b) $e^2 e^3 e$
   (c) $e^{3x}(e^x - e^{2x}) - (e^{2x})^2$

   (d) $e^{3+t}e^{5-2t}$ (e) $\dfrac{e^{2z+3}e^{5-z}}{e^{z+2}}$

   (f) $\sqrt{e^{4z}}$ (g) $(e^{x/2}e^x)^2$

   (h) $\sqrt{e^{2t} + 2e^{2t+1} + e^{2t+2}}$

   (i) $\dfrac{e^t + e^{2t}}{e^{2t} + e^{3t}}$

4. Use a graphical method to solve

   (a) $e^{2x} = 2x + 2$

   (b) $e^{-x} = 2 - x^2$

   (c) $e^x = 2 + \dfrac{x^3}{2}$

5. (a) Sketch the graphs of

   $y = 2 + e^{-x}$ and $y = e^x$

   for $0 \le x \le 2$.

   (b) Use your graphs to solve

   $e^x - e^{-x} = 2$

6. Remove the brackets from the following expressions:

   (a) $(e^{2t})^4$ (b) $e^t(e^t - e^{-t})$

   (c) $e^{2y}(e^{3y} + e^{-2y} + e^y)$ (d) $(1 + e^t)^2$

15 The logarithm function

Objectives

This chapter
- explains the term 'base of a logarithm'
- shows how to calculate the logarithm of a number to any base
- states the laws of logarithms and uses them to simplify expressions
- shows how to solve exponential and logarithmic equations
- defines the logarithm function
- Illustrates graphically the logarithm function

15.1   Introducing logarithms

Given an equation such as $125 = 5^3$, we call 5 the base and 3 the power or index. We can use **logarithms** to write the equation in another form. The logarithm form is

$$\log_5 125 = 3$$

This is read as 'logarithm to the base 5 of 125 is 3'. In general if

$$y = a^x$$

then

$$\log_a y = x$$

KEY POINT

> $y = a^x$ and $\log_a y = x$ are equivalent.

'The logarithm to the base a of y is x' is equivalent to saying that 'y is a to the power x'. The word logarithm is usually shortened to just 'log'.

Worked example

15.1   Write down the logarithm form of the following:

(a) $16 = 4^2$ (b) $8 = 2^3$ (c) $25 = 5^2$

Solution   (a) $16 = 4^2$ may be written as

$$2 = \log_4 16$$

that is, 2 is the logarithm to the base 4 of 16.

(b) $8 = 2^3$ may be expressed as

$$3 = \log_2 8$$

which is read as '3 is the logarithm to the base 2 of 8'.

(c) $25 = 5^2$ may be expressed as

$$2 = \log_5 25$$

that is, 2 is the logarithm to the base 5 of 25.

Given an equation such as $16 = 4^2$ then 'taking logs' to base 4 will result in the logarithmic form $\log_4 16 = 2$.

Worked example

15.2   Write the exponential form of the following:

(a) $\log_2 16 = 4$ (b) $\log_3 27 = 3$ (c) $\log_5 125 = 3$ (d) $\log_{10} 100 = 2$

Solution   (a) Here the base is 2 and so we may write $16 = 2^4$.

(b) The base is 3 and so $27 = 3^3$.

(c) The base is 5 and so $125 = 5^3$.

(d) The base is 10 and so $100 = 10^2$.

Recall that e is the constant 2.71828..., which occurs frequently in natural phenomena

Although the base of a logarithm can be any positive number, the commonly used bases are 10 and e. Logarithms to base 10 are often denoted by 'log' or 'log$_{10}$'; logarithms to base e are denoted by 'ln' or 'log$_e$' and referred to as **natural logarithms**. Most scientific calculators possess 'log' and 'ln' buttons which are used to evaluate logarithms to base 10 and base e.

Worked examples

15.3   Use a scientific calculator to evaluate the following:

(a) $\log 71$ (b) $\ln 3.7$ (c) $\log 0.4615$ (d) $\ln 0.5$ (e) $\ln 1000$

Solution   Using a scientific calculator we obtain

(a) $\log 71 = 1.8513$ (b) $\ln 3.7 = 1.3083$ (c) $\log 0.4615 = -0.3358$

(d) $\ln 0.5 = -0.6931$ (e) $\ln 1000 = 6.9078$
You should ensure that you know how to use your calculator to verify these results.

15.4   Given $10^{0.6990} = 5$ evaluate the following:

(a) $10^{1.6990}$ (b) $\log 5$ (c) $\log 500$

Solution   (a) $10^{1.6990} = 10^1\, 10^{0.6990} = 10(5) = 50$

(b) We are given $10^{0.6990} = 5$ and by taking logs to the base 10 we obtain $0.6990 = \log 5$.

(c) From $10^{0.6990} = 5$ we can see that

$$100(5) = 100(10^{0.6990})$$

and so

$$500 = 10^2.10^{0.6990} = 10^{2.6990}$$

Taking logs to the base 10 gives
$$\log 500 = 2.6990$$

Self-assessment questions 15.1

1. The base of a logarithm is always a positive integer. True or false?

Exercise 15.1

1. Use a scientific calculator to evaluate

(a) $\log 150$ (b) $\ln 150$ (c) $\log 0.316$
(d) $\ln 0.1$

2. Write down the logarithmic form of the following:

(a) $3^8 = 6561$ (b) $6^5 = 7776$
(c) $2^{10} = 1024$ (d) $10^5 = 100000$
(e) $4^7 = 16384$

(f) $\left(\dfrac{1}{2}\right)^5 = 0.03125$

(g) $12^3 = 1728$ (h) $9^4 = 6561$

3. Write the exponential form of the following:

(a) $\log_6 1296 = 4$ (b) $\log_{15} 225 = 2$
(c) $\log_8 512 = 3$ (d) $\log_7 2401 = 4$
(e) $\log_3 243 = 5$ (f) $\log_6 216 = 3$
(g) $\log_{20} 8000 = 3$ (h) $\log_{16} 4096 = 3$
(i) $\log_2 4096 = 12$

15.2   Calculating logarithms to any base

In §15.1 we showed how equations of the form $y = a^x$ could be written in logarithmic form as $\log_a y = x$. The number a, called the **base**, is always positive, that is, $a > 0$. We also introduced the common bases 10 and e. Scientific calculators are programmed with logarithms to base 10 and to base e. Suppose that we wish to calculate logarithms to bases other than 10 or e. For example, in communication theory, logarithms to base 2 are commonly used. To calculate logarithms to base a we use one of the following two formulae.

KEY POINT

$$\log_a X = \frac{\log_{10} X}{\log_{10} a} \qquad \log_a X = \frac{\ln X}{\ln a}$$

Worked example

15.5   Evaluate (a) $\log_6 19$, (b) $\log_7 29$.

Solution   (a) We use the formula

$$\log_a X = \frac{\log_{10} X}{\log_{10} a}$$

Comparing $\log_6 19$ with $\log_a X$ we see $X = 19$ and $a = 6$. So

$$\log_6 19 = \frac{\log_{10} 19}{\log_{10} 6} = \frac{1.2788}{0.7782} = 1.6433$$

We could have equally well used the formula

$$\log_a X = \frac{\ln X}{\ln a}$$

With this formula we obtain

$$\log_6 19 = \frac{\ln 19}{\ln 6} = \frac{2.9444}{1.7918} = 1.6433$$

(b) Comparing $\log_a X$ with $\log_7 29$ we see $X = 29$ and $a = 7$. Hence

$$\log_7 29 = \frac{\log_{10} 29}{\log_{10} 7} = \frac{1.4624}{0.8451} = 1.7304$$

Alternatively we use

$$\log_7 29 = \frac{\ln 29}{\ln 7} = \frac{3.3673}{1.9459} = 1.7304$$

By considering the formula

$$\log_a X = \frac{\log_{10} X}{\log_{10} a}$$

with $X = a$ we obtain

$$\log_a a = \frac{\log_{10} a}{\log_{10} a} = 1$$

KEY POINT

$$\log_a a = 1$$

This same result could be derived by writing the logarithm form of $a = a^1$.

Exercise 15.2

1. Evaluate the following:

(a) $\log_4 6$ (b) $\log_3 10$ (c) $\log_{20} 270$
(d) $\log_5 0.65$ (e) $\log_2 100$ (f) $\log_2 0.03$
(g) $\log_{100} 10$ (h) $\log_7 7$

15.3   Laws of logarithms

Logarithms obey several laws which we now examine. They are introduced via examples.

Worked example

15.6   Evaluate (a) $\log 7$, (b) $\log 12$, (c) $\log 84$ and $\log 7 + \log 12$. Comment on your findings.

Solution   (a) $\log 7 = 0.8451$ (b) $\log 12 = 1.0792$

(c) $\log 84 = 1.9243$, and $\log 7 + \log 12 = 0.8451 + 1.0792 = 1.9243$

We note that $\log 7 + \log 12 = \log 84$.

Examle 15.6 illustrates the first law of logarithms which states:

KEY POINT

$$\log A + \log B = \log AB$$

This law holds true for any base. However, in any one calculation all bases must be the same.

Worked examples

15.7 Simplify to a single log term

(a) $\log 9 + \log x$

(b) $\log t + \log 4t$

(c) $\log 3x^2 + \log 2x$

Solution (a) $\log 9 + \log x = \log 9x$

(b) $\log t + \log 4t = \log(t.4t) = \log 4t^2$

(c) $\log 3x^2 + \log 2x = \log(3x^2.2x) = \log 6x^3$

15.8 Simplify

(a) $\log 7 + \log 3 + \log 2$

(b) $\log 3x + \log x + \log 4x$

Solution (a) We know $\log 7 + \log 3 = \log(7 \times 3) = \log 21$, and so

$$\log 7 + \log 3 + \log 2 = \log 21 + \log 2$$
$$= \log(21 \times 2) = \log 42$$

(b) We have

$$\log 3x + \log x = \log(3x.x) = \log 3x^2$$

and so

$$\log 3x + \log x + \log 4x = \log 3x^2 + \log 4x$$
$$= \log(3x^2.4x) = \log 12x^3$$

We now consider an example which introduces the second law of logarithms.

15.9 (a) Evaluate $\log 12$, $\log 4$ and $\log 3$.

(b) Compare the values of $\log 12 - \log 4$ and $\log 3$.

Solution   (a) $\log 12 = 1.0792$, $\log 4 = 0.6021$, $\log 3 = 0.4771$.

(b) From part (a),

$$\log 12 - \log 4 = 1.0792 - 0.6021 = 0.4771$$

and also

$$\log 3 = 0.4771$$

We note that $\log 12 - \log 4 = \log 3$.

This example illustrates the second law of logarithms, which states:

KEY POINT

$$\log A - \log B = \log\left(\frac{A}{B}\right)$$

Worked examples

15.10   Use the second law of logarithms to simplify the following to a single log term:

(a) $\log 20 - \log 10$ (b) $\log 500 - \log 75$

(c) $\log 4x^3 - \log 2x$ (d) $\log 5y^3 - \log y$

Solution   (a) Using the second law of logarithms we have

$$\log 20 - \log 10 = \log\left(\frac{20}{10}\right) = \log 2$$

(b) $\log 500 - \log 75 = \log\left(\frac{500}{75}\right) = \log\left(\frac{20}{3}\right)$

(c) $\log 4x^3 - \log 2x = \log\left(\frac{4x^3}{2x}\right) = \log 2x^2$

(d) $\log 5y^3 - \log y = \log\left(\frac{5y^3}{y}\right) = \log 5y^2$

15.11   Simplify

(a) $\log 20 + \log 3 - \log 6$

(b) $\log 18 - \log 24 + \log 2$

Solution   (a) Using the first law of logarithms we see that

$$\log 20 + \log 3 = \log 60$$

and so

$$\log 20 + \log 3 - \log 6 = \log 60 - \log 6$$

Using the second law of logarithms we see that

$$\log 60 - \log 6 = \log\left(\frac{60}{6}\right) = \log 10$$

Hence

$$\log 20 + \log 3 - \log 6 = \log 10$$

(b) $\log 18 - \log 24 + \log 2 = \log\left(\dfrac{18}{24}\right) + \log 2$

$$= \log\left(\frac{3}{4}\right) + \log 2$$

$$= \log\left(\frac{3}{4} \times 2\right)$$

$$= \log 1.5$$

15.12   Simplify

(a) $\log 2 + \log 3x - \log 2x$

(b) $\log 5y^2 + \log 4y - \log 10y^2$

Solution   (a) $\log 2 + \log 3x - \log 2x = \log(2 \times 3x) - \log 2x$

$$= \log 6x - \log 2x$$

$$= \log\left(\frac{6x}{2x}\right) = \log 3$$

(b) $\log 5y^2 + \log 4y - \log 10y^2 = \log(5y^2.4y) - \log 10y^2$

$$= \log 20y^3 - \log 10y^2$$

$$= \log\left(\frac{20y^3}{10y^2}\right) = \log 2y$$

We consider a special case of the second law. Consider $\log A - \log A$. This is clearly 0. However, using the second law we may write

$$\log A - \log A = \log\left(\frac{A}{A}\right) = \log 1$$

Thus

KEY POINT

$$\log 1 = 0$$

In any base, the logarithm of 1 equals 0.

Finally we introduce the third law of logarithms.

Worked example

15.13 (a) Evaluate $\log 16$ and $\log 2$.

(b) Compare $\log 16$ and $4 \log 2$.

Solution (a) $\log 16 = 1.204$, $\log 2 = 0.301$.

(b) $\log 16 = 1.204$, $4 \log 2 = 1.204$. Hence we see that $4 \log 2 = \log 16$.

Noting that $16 = 2^4$, example 15.13 suggests the third law of logarithms:

KEY POINT

$$n \log A = \log A^n$$

This law applies if n is integer, fractional, positive or negative.

Worked example

15.14 Write the following as a single logarithmic expression:

(a) $3 \log 2$ (b) $2 \log 3$ (c) $4 \log 3$

Solution (a) $3 \log 2 = \log 2^3 = \log 8$

(b) $2 \log 3 = \log 3^2 = \log 9$

(c) $4 \log 3 = \log 3^4 = \log 81$

15.15 Write as a single log term

(a) $\frac{1}{2} \log 16$ (b) $-\log 4$ (c) $-2 \log 2$ (d) $-\frac{1}{2} \log 0.5$

Solution (a) $\frac{1}{2} \log 16 = \log 16^{1/2} = \log \sqrt{16} = \log 4$

(b) $-\log 4 = -1 . \log 4 = \log 4^{-1} = \log\left(\frac{1}{4}\right) = \log 0.25$

(c) $-2\log 2 = \log 2^{-2} = \log\left(\frac{1}{2^2}\right) = \log\left(\frac{1}{4}\right) = \log 0.25$

(d) $-\frac{1}{2}\log 0.5 = -\frac{1}{2}\log\left(\frac{1}{2}\right) = \log\left(\frac{1}{2}\right)^{-1/2} = \log 2^{1/2} = \log \sqrt{2}$

15.16 Simplify

(a) $3\log x - \log x^2$

(b) $3\log t^3 - 4\log t^2$

(c) $\log Y - 3\log 2Y + 2\log 4Y$

Solution (a) $3\log x - \log x^2 = \log x^3 - \log x^2$

$$= \log\left(\frac{x^3}{x^2}\right)$$

$$= \log x$$

(b) $3\log t^3 - 4\log t^2 = \log(t^3)^3 - \log(t^2)^4$

$$= \log t^9 - \log t^8$$

$$= \log\left(\frac{t^9}{t^8}\right)$$

$$= \log t$$

(c) $\log Y - 3\log 2Y + 2\log 4Y = \log Y - \log(2Y)^3 + \log(4Y)^2$

$$= \log Y - \log 8Y^3 + \log 16Y^2$$

$$= \log\left(\frac{Y.16Y^2}{8Y^3}\right)$$

$$= \log 2$$

15.17 Simplify

(a) $2\log 3x - \frac{1}{2}\log 16x^2$

(b) $\frac{3}{2}\log 4x^2 - \log\left(\frac{1}{x}\right)$

(c) $2\log\left(\frac{2}{x^2}\right) - 3\log\left(\frac{2}{x}\right)$

Solution   (a) $2\log 3x - \dfrac{1}{2}\log 16x^2 = \log(3x)^2 - \log(16x^2)^{1/2}$

$$= \log 9x^2 - \log 4x$$

$$= \log\left(\frac{9x^2}{4x}\right)$$

$$= \log\left(\frac{9x}{4}\right)$$

(b) $\dfrac{3}{2}\log 4x^2 - \log\left(\dfrac{1}{x}\right) = \log(4x^2)^{3/2} - \log(x^{-1})$

$$= \log 8x^3 + \log x$$

$$= \log 8x^4$$

(c) $2\log\left(\dfrac{2}{x^2}\right) - 3\log\left(\dfrac{2}{x}\right) = \log\left(\dfrac{2}{x^2}\right)^2 - \log\left(\dfrac{2}{x}\right)^3$

$$= \log\left(\frac{4}{x^4}\right) - \log\left(\frac{8}{x^3}\right)$$

$$= \log\left(\frac{4/x^4}{8/x^3}\right)$$

$$= \log\left(\frac{1}{2x}\right)$$

Self-assessment questions 15.3

1. State the three laws of logarithms.

Exercise 15.3

1. Write the following as a single log term using the laws of logarithms:

   (a) $\log 5 + \log 9$ (b) $\log 9 - \log 5$
   (c) $\log 5 - \log 9$ (d) $2\log 5 + \log 1$
   (e) $2\log 4 - 3\log 2$ (f) $\log 64 - 2\log 2$
   (g) $3\log 4 + 2\log 1 + \log 27 - 3\log 12$

   (f) $2\log y - \log y^2$

   (g) $3\log\left(\dfrac{1}{t}\right) + \log t^2$

   (h) $4\log\sqrt{x} + 2\log\left(\dfrac{1}{x}\right)$

   (i) $2\log x + 3\log t$ (j) $\log A - \dfrac{1}{2}\log 4A$

2. Simplify the following as much as possible by using the laws of logarithms:

   (a) $\log 4x - \log x$ (b) $\log t^3 + \log t^4$

   (c) $\log 2t - \log\left(\dfrac{t}{4}\right)$

   (d) $\log 2 + \log\left(\dfrac{3}{x}\right) - \log\left(\dfrac{x}{2}\right)$

   (e) $\log\left(\dfrac{t^2}{3}\right) + \log\left(\dfrac{6}{t}\right) - \log\left(\dfrac{1}{t}\right)$

   (k) $\dfrac{\log 9x + \log 3x^2}{3}$

   (l) $\log xy + 2\log\left(\dfrac{x}{y}\right) + 3\log\left(\dfrac{y}{x}\right)$

   (m) $\log\left(\dfrac{A}{B}\right) - \log\left(\dfrac{B}{A}\right)$

   (n) $\log\left(\dfrac{2t}{3}\right) + \dfrac{1}{2}\log 9t - \log\left(\dfrac{1}{t}\right)$

15.4   Solving equations with logarithms

This section illustrates the use of logarithms in solving certain types of equations. For reference we note from §15.1 the equivalence of

$$y = a^x \text{ and } \log_a y = x$$

and from §15.3 the laws of logarithms:

$$\log A + \log B = \log AB$$

$$\log A - \log B = \log\left(\frac{A}{B}\right)$$

$$n \log A = \log A^n$$

Worked examples

15.18   Solve the following equations

(a) $10^x = 59$ (b) $10^x = 0.37$ (c) $e^x = 100$ (d) $e^x = 0.5$

Solution   (a) $10^x = 59$

Taking logs to base 10 gives

$$x = \log 59 = 1.7709$$

(b) $10^x = 0.37$

Taking logs to base 10 gives

$$x = \log 0.37 = -0.4318$$

(c) $e^x = 100$

Taking logs to base e gives

$$x = \ln 100 = 4.6052$$

(d) $e^x = 0.5$

Taking logs to base e we have

$$x = \ln 0.5 = -0.6931$$

15.19   Solve the following equations

(a) $\log x = 1.76$ (b) $\ln x = -0.5$ (c) $\log(3x) = 0.76$

(d) $\ln\left(\frac{x}{2}\right) = 2.6$ (e) $\log(2x - 4) = 1.1$ (f) $\ln(7 - 3x) = 1.75$

Solution   We note that if $\log_a X = n$, then $X = a^n$.

(a) $\log x = 1.76$ and so $x = 10^{1.76}$. Using the 'x^y' button of a scientific calculator we find

$$x = 10^{1.76} = 57.5440$$

(b) $\ln x = -0.5$

$$x = e^{-0.5} = 0.6065$$

(c) $\log 3x = 0.76$

$$3x = 10^{0.76}$$

$$x = \frac{10^{0.76}}{3} = 1.9181$$

(d) $\ln\left(\frac{x}{2}\right) = 2.6$

$$\frac{x}{2} = e^{2.6}$$

$$x = 2e^{2.6} = 26.9275$$

(e) $\log(2x - 4) = 1.1$

$$2x - 4 = 10^{1.1}$$

$$2x = 10^{1.1} + 4$$

$$x = \frac{10^{1.1} + 4}{2} = 8.2946$$

(f) $\ln(7 - 3x) = 1.75$

$$7 - 3x = e^{1.75}$$

$$3x = 7 - e^{1.75}$$

$$x = \frac{7 - e^{1.75}}{3} = 0.4151$$

15.20   Solve the following:

(a) $e^{3+x}.e^x = 1000$ (b) $\ln\left(\frac{x}{3} + 1\right) + \ln\left(\frac{1}{x}\right) = -1$

(c) $3e^{x-1} = 75$ (d) $\log(x + 2) + \log(x - 2) = 1.3$

Solution   (a) Using the laws of logarithms we have

$$e^{3+x}.e^x = e^{3+2x}$$

Hence

$$e^{3+2x} = 1000$$

$$3 + 2x = \ln 1000$$

$$2x = \ln(1000) - 3$$

$$x = \frac{\ln(1000) - 3}{2} = 1.9539$$

(b) Using the laws of logarithms we may write

$$\ln\left(\frac{x}{3}+1\right)+\ln\left(\frac{1}{x}\right)=\ln\left(\frac{x}{3}+1\right)\frac{1}{x}=\ln\left(\frac{1}{3}+\frac{1}{x}\right)$$

So $\ln\left(\frac{1}{3}+\frac{1}{x}\right)=-1$

$$\frac{1}{3}+\frac{1}{x}=e^{-1}$$

$$\frac{1}{x}=e^{-1}-\frac{1}{3}$$

$$\frac{1}{x}=0.0345$$

$$x=28.947$$

(c) $3e^{x-1}=75$

$$e^{x-1}=25$$

$$x-1=\ln 25$$

$$x=\ln 25+1=4.2189$$

(d) Using the first law we may write

$$\log(x+2)+\log(x-2)=\log(x+2)(x-2)=\log(x^2-4)$$

Hence

$$\log(x^2-4)=1.3$$

$$x^2-4=10^{1.3}$$

$$x^2=10^{1.3}+4$$

$$x=\sqrt{10^{1.3}+4}=4.8941$$

Exercise 15.4

1. Solve the following equations, giving your answer to 4 d.p.:

(a) $\log x=0.3940$ (b) $\ln x=0.3940$
(c) $10^y=5.5$ (d) $e^z=500$

(e) $\log(3v)=1.6512$ (f) $\ln\left(\frac{t}{6}\right)=1$

(g) $10^{2r+1}=25$ (h) $e^{(2t-1)/3}=7.6700$
(i) $\log(4b^2)=2.6987$

(j) $\log\left(\frac{6}{2+t}\right)=1.5$

(k) $\ln(2r^3+1)=3.0572$
(l) $\ln(\log t)=-0.3$ (m) $10^{t^2-1}=180$
(n) $10^{3r^4}=170\,000$

(o) $\log(10^t)=1.6$ (p) $\ln(e^x)=20\,000$

2. Solve the following equations:

(a) $e^{3x}.e^{2x}=59$
(b) $10^{3t}.10^{4-t}=27$
(c) $\log(5-t)+\log(5+t)=1.2$
(d) $\log x+\ln x=4$

15.5   Properties and graph of the logarithm function

Values of log x and ln x are given in Table 15.1 and graphs of the functions $y = \log x$ and $y = \ln x$ are illustrated in Figure 15.1. The following properties are noted from the graphs.

  (i) As x increases, the values of log x and ln x increase
  (ii) $\log 1 = \ln 1 = 0$
  (iii) As x approaches 0 the values of log x and ln x increase negatively.
  (iv) When $x < 1$, the values of log x and ln x are negative.

Table 15.1

x	0.01	0.1	0.5	1	2	5	10	100
log x	2	1	−0.30	0	0.30	0.70	1	2
ln x	−4.61	−2.30	−0.69	0	0.69	1.61	2.30	4.61

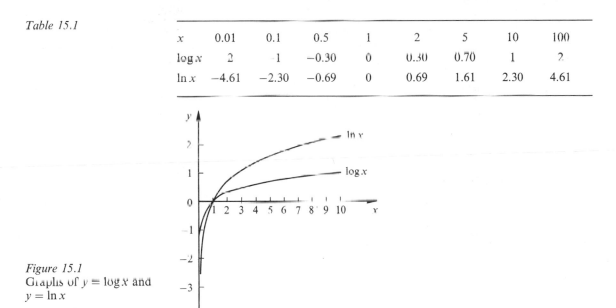

Figure 15.1
Graphs of $y = \log x$ and $y = \ln x$

Self-assessment questions 15.5

1. State two properties which are common to both $y = \log x$ and $y = \ln x$.
2. It is possible to find a value of x such that the value of log x exceeds 100 000 000. True or false?

Exercise 15.5

1. (a) Plot a graph of $y = \log x^2$ for $0 < x \le 10$.
  (b) Plot a graph of $y = \log(1/x)$ for $0 < x \le 10$.

2. (a) Plot a graph of $y = \log x$ for $0 < x \le 3$.

(b) Plot on the same axes $y = 1 - x/3$ for $0 \le x \le 3$.
(c) Use your graphs to find an approximate solution to the equation

$$\log x = 1 - \frac{x}{3}$$

Test and assignment exercises 15

1. Use a scientific calculator to evaluate

   (a) $\log 107\,000$ (b) $\ln 0.0371$ (c) $\log 0.1$
   (d) $\ln 150$

2. Evaluate

   (a) $\log_7 20$ (b) $\log_{16} 100$ (c) $\log_2 60$
   (d) $\log_8 150$ (e) $\log_6 4$

3. Simplify the following expressions to a single log term:

   (a) $\log 7 + \log t$ (b) $\log t - \log x$
   (c) $\log x + 2\log y$ (d) $3\log t + 4\log r$

   (e) $\dfrac{1}{2}\log 9v^4 - 3\log 1$

   (f) $\log(a+b) + \log(a-b)$

   (g) $\dfrac{2}{3}\log x + \dfrac{1}{3}\log xy^3$

4. Solve the following logarithmic equations, giving your answer to 4 d.p.:

   (a) $\log 7x = 2.9$ (b) $\ln 2x = 1.5$

   (c) $\ln x + \ln 2x = 3.6$

   (d) $\ln\left(\dfrac{x^2}{2}\right) - \ln x = 0.7$

   (e) $\log(4t - 3) = 0.9$
   (f) $\log 3t + 3\log t = 2$

   (g) $3\ln\left(\dfrac{2}{x}\right) - \dfrac{1}{2}\ln x = -1$

5. (a) Plot $y = \log x$ for $0 < x \le 3$
   (b) On the same axes plot $y = \sin x$ where x is measured in radians.
   (c) Hence find an approximate solution to $\log x = \sin x$.

16 Angles

Objectives

This chapter

- explains the units 'degree' and 'radian'
- shows how to convert from degrees to radians and from radians to degrees
- shows how to calculate the length of a circular arc
- shows how to calculate the area of a sector

16.1   Measuring angles

We use angles to measure the amount by which a line has been turned. An angle is commonly denoted by θ. In Figure 16.1 the angle between BC and BA is θ. We write $\angle ABC = \theta$. We could also write $\angle CBA = \theta$. The two units for measuring angle are the **degree** and the **radian**. We examine each in turn.

θ is the Greek letter 'theta' commonly used to denote angles

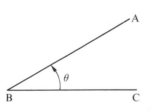

Figure 16.1
The angle ABC is θ

Degrees

A full revolution is 360 degrees, denoted 360°. To put it another way, 1 degree (1°) is $\frac{1}{360}$th of a full revolution. In Figure 16.2, AB is rotated through half a revolution, that is, 180°, to the position AC. CAB is a straight line and so 180° is sometimes called a **straight line angle**.

Figure 16.2
180° is a straight line
angle

Figure 16.3 shows an angle of 90°. This is the angle in a corner. An angle of 90° is called a **right angle**. A triangle containing a right angle is a **right-angled triangle**. The side opposite the right angle is called the **hypotenuse**. An angle between 0° and 90° is an **acute angle**. An angle between 90° and 180° is an **obtuse angle**. An angle greater than 180° is a **reflex angle**.

Figure 16.3
An angle of 90° is a right
angle

Radians

The other unit of angle is the radian and this too is based on examining a circle. Radian is short for 'radius angle'. Consider Figure 16.4 which shows a circle of radius *r*, centre O and an arc AB. We say that the arc AB **subtends** an angle AOB at the centre of the circle. Clearly if AB is short then ∠AOB will be small; if AB is long then ∠AOB will be a large angle. If AB has a length equal to 1 radius then we say that ∠AOB is 1 **radian**.

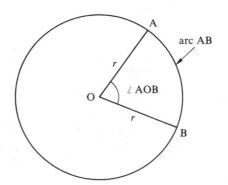

Figure 16.4
Arc AB subtends ∠AOB
at the centre of the circle

KEY POINT

1 radian is the angle subtended at the centre of a circle by an arc length of 1 radius.

We abbreviate 1 radian to 1^c. From the definition we see that an arc length $2r$, that is, 2 radii, subtends an angle of 2^c, an arc length of $3r$ subtends an angle of 3^c and so on. In particular, an arc length of $2\pi r$ subtends an angle of $2\pi^c$. We note that the circumference of the circle is $2\pi r$ and a full circumference subtends a full revolution, that is, $360°$, at the centre. Hence

$$2\pi^c = 360°$$

and so

KEY POINT

$$\pi^c = 180°$$

Your scientific calculator is pre-programmed to give a value for π. Check that you can use it correctly.

Recall that π is an irrational constant whose value is approximately 3.142.

This equation allows us to convert degrees to radians and radians to degrees.

Worked examples

16.1   Convert to degrees

(a) $\dfrac{\pi^c}{2}$ (b) $\dfrac{\pi^c}{4}$ (c) $\dfrac{\pi^c}{3}$ (d) $0.7\pi^c$ (e) 1^c (f) 3^c (g) 1.3^c

Solution   (a) We have

$$\pi^c = 180°$$

and so

$$\frac{\pi^c}{2} = \frac{180°}{2} = 90°$$

Hence $\dfrac{\pi^c}{2}$ is a right angle.

(b) $\pi^c = 180°$

$$\frac{\pi^c}{4} = \frac{180°}{4}$$
$$= 45°$$

(c) $\pi^c = 180°$

$$\frac{\pi^c}{3} = \frac{180°}{3}$$
$$= 60°$$

(d) $\pi^c = 180°$

$$0.7\pi^c = 0.7 \times 180°$$
$$= 126°$$

(e) $\pi^c = 180°$

$1^c = \dfrac{180°}{\pi}$

$= 57.3°$

(g) $1^c = \dfrac{180°}{\pi}$

$1.3^c = 1.3 \times \dfrac{180°}{\pi}$

$= 74.5°$

(f) $\pi^c = 180°$

$1^c = \dfrac{180°}{\pi}$

$3^c = 3 \times \dfrac{180°}{\pi}$

$= 171.9°$

16.2   Convert to radians

(a) 72° (b) 120° (c) 12° (d) 200°

Solution   (a) $180° = \pi^c$

$1° = \dfrac{\pi^c}{180}$

$72° = 72 \times \dfrac{\pi^c}{180}$

$= \dfrac{2\pi^c}{5} = 1.26^c$

(b) $180° = \pi^c$

$1° = \dfrac{\pi^c}{180}$

$120° = 120 \times \dfrac{\pi^c}{180}$

$= \dfrac{2\pi^c}{3} = 2.09^c$

(c) $180° = \pi^c$

$1° = \dfrac{\pi^c}{180}$

$12° = 12 \times \dfrac{\pi^c}{180}$

$= \dfrac{\pi^c}{15} = 0.21^c$

(d) $180° = \pi^c$

$1° = \dfrac{\pi^c}{180}$

$200° = 200 \times \dfrac{\pi^c}{180}$

$= \dfrac{10\pi^c}{9} = 3.49^c$

Self-assessment questions 16.1

1. Define the units of angle: degree and radian.
2. How are the degree and the radian related?

Exercise 16.1

1. Convert to radians

   (a) 240° (b) 300° (c) 400° (d) 37°
   (e) 1000°

2. Convert to degrees

   (a) $\dfrac{\pi^c}{10}$ (b) $\dfrac{2\pi^c}{9}$ (c) 2^c (d) 3.46^c (e) 1.75^c

16.2   **Length of an arc**

In §16.1 we saw the connection between the length of an arc and the angle it subtends at the centre. Knowing the angle subtended by an arc we are able to calculate the length of the arc.

An arc of length 1 × radius subtends an angle of 1^c. An arc of length 2 × radius subtends an angle of 2^c. Clearly an arc of length θ × radius subtends an angle of θ^c.

KEY POINT

An angle θ^c is subtended by an arc of length $r\theta$. Note that θ must be in radians. Hence arc length $= r\theta$

Worked examples

16.3   A circle has a radius of 8 cm. An angle of 1.4^c is subtended at the centre by an arc. Calculate the length of the arc.

Solution   An angle of 1^c is subtended by an arc of length 1 radius. So an angle of 1.4^c is subtended by an arc of length 1.4 × radius. Now

$$1.4 \times \text{radius} = 1.4 \times 8 = 11.2$$

The length of the arc is 11.2 cm.

16.4   A circle has radius 6 cm. An arc, AB, of the circle measures 9 cm. Calculate the angle subtended by AB at the centre of the circle.

Solution   Using arc length $= r\theta$ we have

$$9 = 6\theta$$
$$\theta = \frac{9}{6} = 1.5^c$$

Hence AB subtends an angle of 1.5^c at the centre.

16.5   A circle has a radius of 18 cm. An arc AB subtends an angle of 142° at the centre of the circle. Calculate the length of the arc AB.

Solution   We convert 142° to radians,

$$180° = \pi^c$$
$$142° = \frac{142 \times \pi^c}{180}$$

An angle of $\dfrac{142\pi}{180}$ is subtended by an arc of length

$\dfrac{142\pi}{180} \times$ radius.

$$\dfrac{142\pi}{180} \times \text{radius} = \dfrac{142\pi}{180} \times 18 = 44.61$$

The length of the arc is 44.61 cm.

Exercise 16.2

1. A circle has a radius of 15 cm. Calculate the length of arc which subtends an angle of

    (a) 2^c (b) 3^c (c) 1.2^c (d) $100°$ (e) $217°$

2. A circle has radius 12 cm. Calculate the angle subtended at the centre by an arc of length

    (a) 12 cm (b) 6 cm (c) 24 cm (d) 15 cm (e) 2 cm

<table>
<tr><td>16.3</td><td>

Area of a sector

</td></tr>
</table>

In Figure 16.5 the shaded area is a **sector**. Suppose the radius of the circle is r and the angle subtended at the centre is θ^c. We can calculate the area of the sector in terms of r and θ. When the angle subtended is $2\pi^c$, the sector occupies the full circle and so has an area of πr^2. When the angle subtended is π^c, the sector is a semi-circle and so the sector has an area of $\pi r^2/2$. Thus, if the angle subtended is 1^c the area of the sector is $\pi r^2/2\pi = r^2/2$. Finally, if the angle subtended is θ^c, the area of the sector is $r^2\theta/2$.

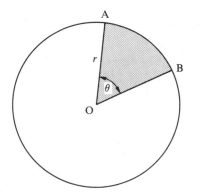

Figure 16.5
The sector AOB is shaded

KEY POINT

$$\text{Area of sector} = \frac{r^2\theta}{2}$$

Note that this formula is only valid when the angle θ is measured in radians.

Worked examples

16.6   A circle has a radius of 22 cm. A sector subtends an angle of 1.2^c. Calculate the area of the sector.

Solution   Here $r = 22$ and $\theta = 1.2$. Hence

$$\text{area of sector} = \frac{(22)^2(1.2)}{2} = 290.4$$

The area of the sector is 290.4 cm².

16.7   A circle has a radius of 10 cm. The area of a sector is 170 cm². Calculate the angle at the centre.

Solution

$$\text{area of sector} = \frac{r^2\theta}{2}$$

Substituting the values given we obtain

$$170 = \frac{(10)^2\theta}{2}$$
$$\theta = \frac{2 \times 170}{10^2}$$
$$= 3.4$$

The angle is 3.4^c.

Self-assessment questions 16.3

1. Explain what is meant by a sector of a circle. Describe how to calculate the area of a sector.

Exercise 16.3

1. A circle has radius of 9 cm. Calculate the area of the sectors which subtend angles of

(a) 1.5^c (b) 2^c (c) $100°$ (d) $215°$

2. A circle has a radius of 16 cm. Calculate the angle at the centre when a sector has an area of

(a) 100 cm² (b) 5 cm² (c) 520 cm²

Test and assignment exercises 16

1. Convert to radians

   (a) 73° (b) 196° (c) 1000°

2. Convert to degrees

   (a) 1.75^c (b) 0.004^c (c) 7.9^c

3. A circle, centre O, has radius 22 cm. An arc AB has length 35 cm.

   (a) Calculate the angle at O subtended by AB.
   (b) Calculate the area of the sector AOB.

4. A circle, centre O, has radius 42 cm. Arc AB subtends an angle of 235° at O.

   (a) Calculate the length of the arc AB.
   (b) Calculate the area of the sector AOB.

5. A circle, centre O, has radius 20 cm. The area of sector AOB is 320 cm².

   (a) Calculate the angle at O subtended by the arc AB.
   (b) Calculate the length of the arc AB.

6. A sector AOB has an area of 250 cm² and the angle at the centre, O, is 75°.

   (a) Calculate the area of the circle.
   (b) Calculate the radius of the circle.

17 Trigonometry

Objectives

This chapter

- defines the trigonometrical ratios sine, cosine and tangent
- illustrates the graphs of the trigonometrical functions
- explains the inverse of the trigonometrical functions
- shows how to solve trigonometrical equations

17.1 The trigonometrical ratios

Consider the right-angled triangle ABC as shown in Figure 17.1. There is a right-angle at B and the angle A is θ. The hypotenuse is the side opposite the right-angle, that is, AC. The side opposite θ is BC. We can calculate BC/AC. This is known as the **sine** of θ, or simply as $\sin\theta$. Similarly, the side adjacent to θ is AB and we call AB/AC the **cosine** of θ, written $\cos\theta$. Finally BC/AB is known as the **tangent** of θ, written $\tan\theta$. Sine, cosine and tangent are known as the **trigonometrical ratios**.

Figure 17.1
A right-angled triangle
ABC

KEY POINT

$$\sin \theta = \frac{\text{side opposite to } \theta}{\text{hypotenuse}} = \frac{BC}{AC}$$

$$\cos \theta = \frac{\text{side adjacent to } \theta}{\text{hypotenuse}} = \frac{AB}{AC}$$

$$\tan \theta = \frac{\text{side opposite to } \theta}{\text{side adjacent to } \theta} = \frac{BC}{AB}$$

Note that all the trigonometrical ratios are defined as the ratio of two lengths and so they themselves have no units. Also, since the hypotenuse is always the longest side of a triangle, then $\sin \theta$ and $\cos \theta$ can never be greater than 1.

Worked example

17.1   Calculate $\sin \theta, \cos \theta$ and $\tan \theta$ for $\triangle ABC$ as shown in Figure 17.2.

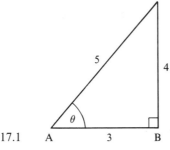

Figure 17.2
$\triangle ABC$ for Example 17.1

Solution

$$\sin \theta = \frac{BC}{AC} = \frac{4}{5} = 0.8$$
$$\cos \theta = \frac{AB}{AC} = \frac{3}{5} = 0.6$$
$$\tan \theta = \frac{BC}{AB} = \frac{4}{3} = 1.3333$$

It is not always necessary to draw triangles when calculating the trigonometrical ratios. Values of the trigonometrical ratios can be found using a scientific calculator. Your calculator must be instructed to work in either radians or degrees. Usually the MODE button is used to select the required units.

Worked examples

17.2 Use a scientific calculator to evaluate

(a) $\sin 65°$ (b) $\cos 17°$ (c) $\tan 50°$ (d) $\sin 1^c$ (e) $\cos 1.5^c$ (f) $\tan 0.5^c$

Solution For (a), (b) and (c) ensure your calculator is set to DEGREES, and not to RADIANS or GRADS.

(a) $\sin 65° = 0.9063$ (b) $\cos 17° = 0.9563$ (c) $\tan 50° = 1.1918$

Now, change the MODE of your calculator in order to work in radians. Check that

(d) $\sin 1^c = 0.8415$ (e) $\cos 1.5^c = 0.0707$ (f) $\tan 0.5^c = 0.5463$

17.3 Prove

$$\frac{\sin \theta}{\cos \theta} = \tan \theta$$

Solution From Figure 17.1, we have

$$\sin \theta = \frac{BC}{AC} \qquad \cos \theta = \frac{AB}{AC}$$

and so

$$\frac{\sin \theta}{\cos \theta} = \frac{BC/AC}{AB/AC} = \frac{BC}{AC} \times \frac{AC}{AB} = \frac{BC}{AB}$$

But

$$\tan \theta = \frac{BC}{AB}$$

and so

$$\frac{\sin \theta}{\cos \theta} = \tan \theta$$

Self-assessment questions 17.1

1. Define the trigonometrical ratios sine, cosine and tangent with reference to a right-angled triangle.
2. Explain why the trigonometrical ratios have no units.

Exercise 17.1

1. Evaluate

(a) $\tan 30°$ (b) $\sin 20°$ (c) $\cos 75°$

2. (a) Evaluate $\sin 70°$ and $\cos 20°$.
   (b) Evaluate $\sin 25°$ and $\cos 65°$.
   (c) Prove $\sin \theta = \cos(90° - \theta)$.

3. A right-angled triangle ABC has AB = 3 cm, BC = 3 cm and AC = $\sqrt{18}$ cm. $\angle CAB = \theta$. Without using a calculator find (a) $\sin \theta$, (b) $\cos \theta$, (c) $\tan \theta$.

17.2   Extended definition of the trigonometrical ratios

In §17.1 we used a right-angled triangle in order to define the three trigonometrical ratios. The angle θ is thus limited to a maximum value of 90°. To give meaning to the trigonometrical ratios of angles greater than 90° we introduce an extended definition.

The x and y axes divide the plane into four quadrants as shown in Figure 17.3. The origin is O. We consider an arm OC which can rotate into any of the quadrants. In Figure 17.3, the arm is in quadrant 1. We measure the anticlockwise angle from the positive x axis to the arm and call this angle θ. When the arm is in quadrant 1, then $0° \le \theta \le 90°$, when in the second quadrant $90° \le \theta \le 180°$, when in the third quadrant $180° \le \theta \le 270°$ and when in the fourth quadrant $270° \le \theta \le 360°$. Note that by rotating the arm clockwise from the x axis, the angle so formed is negative. We consider the x and y projections of the arm OC. These are illustrated in Figure 17.4. We label the x projection OB and the y projection OA. One or both of these projections can be negative, depending upon the position of the arm OC. However, the arm itself is always considered to be positive. The trigonometrical ratios are now defined as

$$\sin\theta = \frac{y \text{ projection of arm OC}}{\text{OC}} = \frac{\text{OA}}{\text{OC}}$$

$$\cos\theta = \frac{x \text{ projection of arm OC}}{\text{OC}} = \frac{\text{OB}}{\text{OC}}$$

$$\tan\theta = \frac{y \text{ projection of arm OC}}{x \text{ projection of arm OC}} = \frac{\text{OA}}{\text{OB}}$$

Figure 17.3
The plane is divided into four quadrants

Figure 17.4
The x projection of OC is OB; the y projection of OC is OA

Note that the extended definition is in terms of projections and so θ is not limited to a maximum value of 90°.

Consider the arm OC in the first quadrant as shown in Figure 17.4. From the right-angled triangle OCB we have

$$\sin\theta = \frac{BC}{OC} \qquad \cos\theta = \frac{OB}{OC} \qquad \tan\theta = \frac{BC}{OB}$$

Alternatively, using the extended definition we could also write

$$\sin\theta = \frac{OA}{OC} \qquad \cos\theta = \frac{OB}{OC} \qquad \tan\theta = \frac{OA}{OB}$$

Noting that OA = BC, we see that the two definitions are in agreement when $0° \le \theta \le 90°$.

We now consider the arm in the second quadrant as shown in Figure 17.5. The x projection of OC is onto the negative part of the x axis; the y projection of OC is onto the positive part of the y axis. Hence $\sin\theta$ is positive, whereas $\cos\theta$ and $\tan\theta$ are negative.

Figure 17.5
When OC is in the second quadrant the x projection is negative

Figure 17.6

The trigonometrical ratios of θ may be related to the trigonometrical ratios of an angle in the first quadrant. Clearly $\angle COB = 180° - \theta$. We draw such an angle in the first quadrant by introducing arm OC' as shown in Figure 17.6. The projections of OC' are OB' and OA. In the first quadrant, using projections we have

$$\sin(180° - \theta) = \frac{OA}{OC'}$$

In the second quadrant

$$\sin\theta = \frac{OA}{OC}$$

Because OC = OC' we see

$$\sin\theta = \sin(180° - \theta)$$

Similarly in the first quadrant

$$\cos(180° - \theta) = \frac{OB'}{OC'}$$

and in the second quadrant

$$\cos\theta = \frac{OB}{OC}$$

Since $OB = -OB'$ we have

$$\cos(180° - \theta) = -\cos\theta$$

A similar analysis shows

$$\tan(180° - \theta) = -\tan\theta$$

These results may be summarized as follows:

KEY POINT

$$\sin(180° - \theta) = \sin\theta$$
$$\cos(180° - \theta) = -\cos\theta$$
$$\tan(180° - \theta) = -\tan\theta$$

These results are illustrated in the following example.

Worked example

17.4   Evaluate

(a) $\sin 40°$ and $\sin 140°$

(b) $\cos 70°$ and $\cos 110°$

(c) $\tan 30°$ and $\tan 150°$

Solution   (a) $\sin 40° = 0.6428$ and $\sin 140° = 0.6428$, hence $\sin 40° = \sin 140°$.

(b) $\cos 70° = 0.3420$ and $\cos 110° = -0.3420$, hence
$\cos 110° = -\cos 70°$.

(c) $\tan 30° = 0.5774$ and $\tan 150° = -0.5774$, hence
$\tan 150° = -\tan 30°$.

When the arm is in the third quadrant as shown in Figure 17.7, the x and y projections are both negative. Hence for $180° < \theta < 270°$ $\sin\theta$ and $\cos\theta$ are negative and $\tan\theta$ is positive. We now relate $\sin\theta$, $\cos\theta$ and $\tan\theta$ to an angle in the first quadrant. We see $\angle BOC = \theta - 180°$ and draw such an angle in the first quadrant using arm OC'. This is illustrated in Figure 17.8. By considering the projections in the same way as before, it is possible to show that:

KEY POINT

$$\sin(\theta - 180°) = -\sin\theta$$
$$\cos(\theta - 180°) = -\cos\theta$$
$$\tan(\theta - 180°) = \tan\theta$$

Figure 17.7
Both x and y projections
are negative

Figure 17.9

Worked example

17.5   Evaluate

(a) $\sin 30°$ and $\sin 210°$

(b) $\cos 40°$ and $\cos 220°$

(c) $\tan 50°$ and $\tan 230°$

Solution   (a) $\sin 30° = 0.5$ and $\sin 210° = -0.5$. Note that $\sin 30° = -\sin 210°$.

(b) $\cos 40° = 0.7660$ and $\cos 220° = -0.7660$. Note that $\cos 40° = -\cos 220°$.

(c) $\tan 50° = 1.1918$ and $\tan 230° = 1.1918$. Note that $\tan 50° = \tan 230°$.

Finally the arm OC is rotated into the fourth quadrant as shown in Figure 17.9. The x projection is positive, the y projection is negative and so $\sin\theta$ and $\tan\theta$ are negative and $\cos\theta$ is positive. Clearly $\angle$BOC $= 360° - \theta$ and such an angle is drawn in the first quadrant using arm OC' as shown in Figure 17.10.

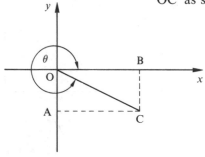

Figure 17.9
The x projection is positive and the y projection is negative

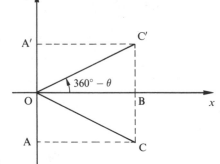

Figure 17.10

Again by considering projections we can show

KEY POINT

$$\sin(360° - \theta) = -\sin\theta$$
$$\cos(360° - \theta) = \cos\theta$$
$$\tan(360° - \theta) = -\tan\theta$$

Worked example

17.6   Evaluate

(a) $\sin 70°$ and $\sin 290°$

(b) $\cos 40°$ and $\cos 320°$

(c) $\tan 20°$ and $\tan 340°$

Solution   (a) $\sin 70° = 0.9397$, $\sin 290° = -0.9397$ and so $\sin 290° = -\sin 70°$.

(b) $\cos 40° = 0.7660$, $\cos 320° = 0.7660$ and so $\cos 40° = \cos 320°$.

(c) $\tan 20° = 0.3640$, $\tan 340° = -0.3640$ and so $\tan 340° = -\tan 20°$.

By looking at $\sin\theta, \cos\theta$ and $\tan\theta$ in the four quadrants we see that:

KEY POINT

	Quadrant 1	Quadrant 2	Quadrant 3	Quadrant 4
$\sin\theta$	positive	positive	negative	negative
$\cos\theta$	positive	negative	negative	positive
$\tan\theta$	positive	negative	positive	negative

We have examined the trigonometrical ratios as θ varies from the first quadrant to the fourth quadrant, that is, from $0°$ to $360°$. It is possible for θ to have values outside the range $0°$ to $360°$. Adding or subtracting $360°$ to an angle is equivalent to rotating the arm through a complete revolution. This will leave its position unchanged. Hence adding or subtracting $360°$ to an angle will leave the trigonometrical ratios unaltered. We state this mathematically as

$$\sin\theta = \sin(\theta + 360°) = \sin(\theta - 360°)$$
$$\cos\theta = \cos(\theta + 360°) = \cos(\theta - 360°)$$
$$\tan\theta = \tan(\theta + 360°) = \tan(\theta - 360°)$$

Worked example

17.7   Evaluate

(a) $\sin 50°$ and $\sin 410°$

(b) $\cos(-110°)$ and $\cos 250°$

(c) $\tan 920°$ and $\tan 200°$

Solution   (a) Using a scientific calculator we have $\sin 50° = 0.7660$. We note that $410° = 50° + 360°$ and hence $\sin 410° = \sin 50° = 0.7660$.

(b) Using a calculator we have $\cos 250° = -0.3420$. Subtracting $360°$ from $250°$ gives $-110°$ and so $\cos(-110°) = \cos 250° = -0.3420$.

(c) Using a calculator we see $\tan 200° = 0.3640$. Now

$$\tan 200° = \tan(200° + 360°) = \tan 560°$$
$$= \tan(560° + 360°) = \tan 920°$$

Hence $\tan 920° = 0.3640$.

Self-assessment questions 17.2

1. State the sign of $\sin\theta$, $\cos\theta$ and $\tan\theta$ in each of the four quadrants.

Exercise 17.2

1. An angle α is such that $\sin\alpha > 0$ and $\cos\alpha < 0$. In which quadrant does α lie?

2. An angle β is such that $\sin\beta < 0$. In which quadrants is it possible for β to lie?

3. In each case state value(s) of θ in the range $0°$ to $360°$ so that the following are true.

   (a) $\sin\theta = \sin 50°$ (b) $\cos\theta = -\cos 40°$
   (c) $\tan\theta = \tan 20°$ (d) $\sin\theta = -\sin 70°$
   (e) $\cos\theta = \cos 10°$ (f) $\tan\theta = -\tan 80°$
   (g) $\sin\theta = \sin 0°$ (h) $\cos\theta = \cos 90°$

17.3	**Graphs of $\sin\theta$, $\cos\theta$ and $\tan\theta$**

The trigonometrical functions are defined by

$$y = \sin\theta \qquad y = \cos\theta \qquad y = \tan\theta$$

The independent variable is θ and the dependent variable is y. By drawing up a table of values it is possible to draw the graphs of the trigonometrical functions. Table 17.1 gives values of θ and corresponding values of $\sin\theta$, $\cos\theta$ and $\tan\theta$. Note that $\tan\theta$ is not defined when $\theta = 90°$, $270°$, $450°$, ... Recalling that $\tan\theta = \sin\theta/\cos\theta$ then whenever $\cos\theta$ is 0, $\tan\theta$ will not be defined. The graphs of $y = \sin\theta$, $y = \cos\theta$ and

Table 17.1
Values of θ and corresponding values of $\sin\theta$, $\cos\theta$ and $\tan\theta$.

θ	0	30	60	90	120	150	180	210
$\sin\theta$	0	0.5000	0.8660	1	0.8660	0.5000	0	−0.5000
$\cos\theta$	1	0.8660	0.5000	0	−0.5000	−0.8660	−1	−0.8660
$\tan\theta$	0	0.5774	1.7321		−1.7321	−0.5774	0	0.5774

θ	240	270	300	330	360	390	420	450
$\sin\theta$	−0.8660	−1	−0.8660	−0.5000	0	0.5000	0.8660	1
$\cos\theta$	−0.5000	0	0.5000	0.8660	1	0.8660	0.5000	0
$\tan\theta$	1.7321		−1.7321	−0.5774	0	0.5774	1.7321	

θ	480	510	540	570	600	630	660	690	720
$\sin\theta$	0.8660	0.5000	0	−0.5000	−0.8660	−1	−0.8660	−0.5000	0
$\cos\theta$	−0.5000	−0.8660	−1	−0.8660	−0.5000	0	0.5000	0.8660	1
$\tan\theta$	−1.7321	−0.5774	0	0.5774	1.7321		−1.7321	−0.5774	0

$y = \tan \theta$ are shown in Figures 17.11, 17.12 and 17.13 respectively. The following properties are noted from the graphs.

(i) sin θ and cos θ repeat the same shape every 360°.
(ii) tan θ repeats the same shape every 180°.
(iii) sin θ and cos θ have a maximum value of 1 and a minimum value of -1.
(iv) tan θ has no maximum value and no minimum value.

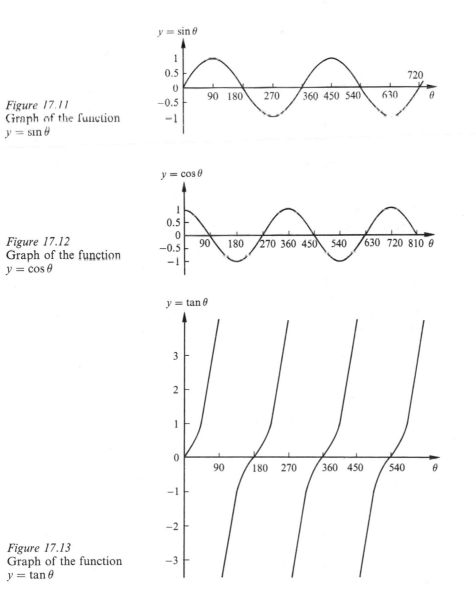

Figure 17.11
Graph of the function
$y = \sin \theta$

Figure 17.12
Graph of the function
$y = \cos \theta$

Figure 17.13
Graph of the function
$y = \tan \theta$

Worked examples

17.8   Draw $y = \sin 2\theta$ for $0° \le \theta \le 180°$.

Solution   Values of θ and $\sin 2\theta$ are given in Table 17.2. A graph of $y = \sin 2\theta$ is shown in Figure 17.14.

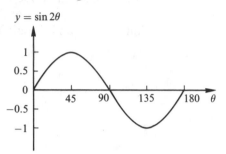

Figure 17.14
A graph of $y = \sin 2\theta$

Table 17.2
Values of θ and corresponding values of $\sin 2\theta$

θ	0	15	30	45	60	75	90	105
2θ	0	30	60	90	120	150	180	210
$\sin 2\theta$	0	0.5000	0.8660	1	0.8660	0.5000	0	−0.5000

θ	120	135	150	165	180
2θ	240	270	300	330	360
$\sin 2\theta$	−0.8660	−1	−0.8660	−0.5000	0

17.9   Draw a graph of $y = \cos(\theta + 30°)$ for $0° \le \theta \le 360°$.

Solution   A table of values is drawn up in Table 17.3. Figure 17.15 shows a graph of $y = \cos(\theta + 30°)$.

Table 17.3
Values of θ and corresponding values of $\cos(\theta + 30°)$

θ	0	30	60	90	120	150	180	210
$\theta + 30$	30	60	90	120	150	180	210	240
$\cos(\theta + 30)$	0.8660	0.5000	0	−0.5000	−0.8660	−1	−0.8660	−0.5000

θ	240	270	300	330	360
$\theta + 30$	270	300	330	360	390
$\cos(\theta + 30)$	0	0.5000	0.8660	1	0.8660

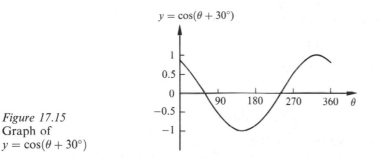

Figure 17.15
Graph of
$y = \cos(\theta + 30°)$

Self-assessment questions 17.3

1. State two properties which $y = \sin\theta$ and $y = \cos\theta$ have in common.
2. State three properties which distinguish $y = \tan\theta$ from either $y = \sin\theta$ or $y = \cos\theta$.

Exercise 17.3

1. Draw the following graphs:

(a) $y = \sin(\theta + 45°)$ for $0° \leq \theta \leq 360°$

(b) $y = 3\cos\left(\dfrac{\theta}{2}\right)$ for $0° \leq \theta \leq 720°$

(c) $y = 2\tan(\theta + 60°)$ for $0° \leq \theta \leq 360°$

17.4   Inverse of the trigonometrical functions

Given an angle θ, we can use a scientific calculator to find $\sin\theta$, $\cos\theta$ and $\tan\theta$. Often we require to reverse the process, that is, given a value of $\sin\theta$, $\cos\theta$ or $\tan\theta$ we need to find the corresponding values of θ.

The superscript −1 does not denote a power. It is a notation for the inverse of the trigonometrical functions

If

$$\sin\theta = x$$

we write

$$\theta = \sin^{-1} x$$

and this is read as 'θ is the inverse sine of x'. This is the same as saying that θ is the angle whose sine is x. Similarly, $\cos^{-1} y$ is the angle whose cosine is y and $\tan^{-1} z$ is the angle whose tangent is z.

Most scientific calculators have inverse sine, inverse cosine and inverse tangent values programmed in. The buttons are usually denoted as $\sin^{-1}$, $\cos^{-1}$ and $\tan^{-1}$.

Worked example

17.10 Use a scientific calculator to evaluate

(a) $\sin^{-1} 0.5$ (b) $\cos^{-1} 0.3$ (c) $\tan^{-1} 2$

Solution (a) Using the $\sin^{-1}$ button we have

$$\sin^{-1} 0.5 = 30°$$

This is another way of saying that $\sin 30° = 0.5$.

(b) Using a scientific calculator we see

$$\cos^{-1} 0.3 = 72.5°$$

that is,

$$\cos 72.5° = 0.3$$

(c) Using a calculator we see $\tan^{-1} 2 = 63.4°$ and hence $\tan 63.4° = 2$.

From Worked Example 17.10(a) we found $\sin^{-1} 0.5 = 30°$, that is, $\sin 30° = 0.5$. However, $\sin 150° = 0.5$ and so it is also true that $\sin^{-1} 0.5 = 150°$. Hence the inverse sine of a number can yield more than one answer. This is also true for inverse cosine and inverse tangent. Scientific calculators will simply give one value. The other values must be deduced from knowledge of the functions $y = \sin \theta$, $y = \cos \theta$ and $y = \tan \theta$.

Worked examples

17.11 Find all values of $\sin^{-1} 0.4$ in the range $0°$ to $360°$.

Solution Let $\theta = \sin^{-1} 0.4$, that is, $\sin \theta = 0.4$. Since $\sin \theta$ is positive then one value is in the first quadrant and another value is in the second quadrant. Using a scientific calculator we see

The formula
$\sin(180° - \theta) = \sin \theta$
was derived on p. 154

$$\theta = \sin^{-1} 0.4 = 23.6°$$

This is the value in the first quadrant. The value in the second quadrant is found using $\sin \theta = \sin(180° - \theta)$. Hence the value in the second quadrant is $180° - 23.6° = 156.4°$.

17.12 Find all values of θ in the range $0°$ to $360°$ such that
(a) $\theta = \cos^{-1}(-0.5)$, (b) $\theta = \tan^{-1} 1$.

Solution (a) We have $\theta = \cos^{-1}(-0.5)$, that is, $\cos \theta = -0.5$. As $\cos \theta$ is negative then θ must be in the second and third quadrants. Using a calculator we find $\cos^{-1}(-0.5) = 120°$. This is the value of θ in the second quadrant. We now seek the value of θ in the third quadrant.

In the third quadrant we have seen (p.155) that

$$\cos(\theta - 180°) = -\cos\theta$$

We are given $\cos\theta = -0.5$

So

$$\cos(\theta - 180°) = 0.5$$

Now, since θ is in the third quadrant, $\theta - 180°$ must be an acute angle whose cosine equals 0.5.

that is

$$\theta - 180° = 60°$$
$$\theta - 240°$$

This is the value of θ in the third quadrant. The required solutions are thus $\theta = 120°, 240°$.

(b) We are given $\theta = \tan^{-1} 1$, that is, $\tan\theta = 1$. Since $\tan\theta$ is positive, there is a value of θ in the first and third quadrants. Using a calculator we find $\tan^{-1} 1 = 45°$. This is the value of θ in the first quadrant.
Now we consider the third quadrant. Here we know (p.155) that

$$\tan(\theta - 180°) = \tan\theta$$

We are given $\tan\theta = 1$

So $\tan(\theta - 180°) = 1$

Now $\theta - 180°$ must be an acute angle with tangent equal to 1

that is $\theta - 180° = 45°$

$$\theta = 225°$$

The required values of θ are 45° and 225°.

17.13 Find all values of θ in the range 0° to 360° such that $\theta = \sin^{-1}(-0.5)$.

Solution We know $\sin\theta = -0.5$ and so θ must be in the third and fourth quadrants. Using a calculator we find

$$\sin^{-1}(-0.5) = -30°$$

We require solutions in the range 0° to 360°. Recall that adding 360° to an angle leaves the values of the trigonometrical ratios unaltered. So

$$\sin^{-1}(-0.5) = -30° + 360° = 330°$$

We have found the value of θ in the fourth quadrant.

We now seek the value in the third quadrant. Here we know (p.155) that

$$\sin(\theta - 180°) = -\sin\theta$$

We are given $\sin\theta = -0.5$

Therefore $\sin(\theta - 180°) = 0.5$

Now $\theta - 180°$ must be an acute angle with sine equal to 0.5

That is, $\theta - 180° = 30°$

$$\theta = 210°$$

The required values of θ are 210° and 330°.

Self-assessment questions 17.4

1. Can you explain why the inverses of the trigonometrical functions have several values for a single value of θ?

Exercise 17.4

1. Find all values in the range 0° to 360° of

   (a) $\sin^{-1} 0.9$ (b) $\cos^{-1} 0.45$ (c) $\tan^{-1} 1.3$
   (d) $\sin^{-1}(-0.6)$ (e) $\cos^{-1}(-0.75)$
   (f) $\tan^{-1}(-0.3)$ (g) $\sin^{-1} 1$ (h) $\cos^{-1}(-1)$

17.5   Solutions of trigonometrical equations

Trigonometrical equations are equations involving the trigonometrical ratios. This section shows how to solve trigonometrical equations.

Worked examples

17.14 Find all values of θ in the range 0° to 360° such that $\sin 2\theta = 0.5$.

Solution We make the substitution $z = 2\theta$. As θ varies from 0° to 360° then z varies from 0° to 720°. Hence the problem as given is equivalent to finding all values of z in the range 0° to 720° such that $\sin z = 0.5$.

$$\sin z = 0.5$$
$$z = \sin^{-1} 0.5$$
$$= 30°$$

Also

$$\sin(180° - 30°) = \sin 30°$$
$$\sin 150° = \sin 30° = 0.5$$
$$\sin^{-1} 0.5 = 150°$$

The values of z in the range 0° to 360° are 30° and 150°. Recalling that adding 360° to an angle leaves its trigonometrical ratios unaltered we see that $30° + 360° = 390°$ and $150° + 360° = 510°$ are values of z in the range 360° to 720°. The solutions for z are thus

$$z = 30°, 150°, 390°, 510°$$

and hence the required values of θ are given by

$$\theta = \frac{z}{2} - 15°, 75°, 195°, 255°$$

17.15   Solve $\cos 3\theta = -0.5$ for $0° \leq \theta \leq 360°$.

Solution   We substitute $z = 3\theta$. As θ varies from 0° to 360° then z varies from 0° to 1080°. Hence the problem is to find values of z in the range 0° to 1080° such that $\cos z = -0.5$.

We begin by finding values of z in the range 0° to 360°. Using Worked Example 17.12(a) we see that

$$z = 120°, 240°$$

To find values of z in the range 360° to 720° we add 360° to each of these solutions. This gives

$$z = 120° + 360° = 480° \quad \text{and} \quad z = 240° + 360° = 600°$$

To find values of z in the range 720° to 1080° we add 360° to these solutions. This gives

$$z = 840°, 960°$$

Hence

$$z = 120°, 240°, 480°, 600°, 840°, 960°$$

Finally, using $\theta = z/3$ we obtain values of θ:

$$\theta = \frac{z}{3} = 40°, 80°, 160°, 200°, 280°, 320°$$

17.16   Solve

$$\tan(2\theta + 20°) = 0.3 \qquad 0° \leq \theta \leq 360°$$

Solution   We substitute $z = 2\theta + 20°$. As θ varies from 0° to 360° then z varies from 20° to 740°.

Firstly we solve

$$\tan z = 0.3 \qquad 0° \leq z \leq 360°$$

This leads to $z = 16.7°, 196.7°$. Values of z in the range $360°$ to $720°$ are

$$z = 16.7° + 360°, 196.7° + 360°$$
$$= 376.7°, 556.7°$$

By adding a further $360°$ values of z in the range $720°$ to $1080°$ are found. These are

$$z = 736.7°, 916.7°$$

Hence values of z in the range $0°$ to $1080°$ are

$$z = 16.7°, 196.7°, 376.7°, 556.7°, 736.7°, 916.7°$$

Values of z in the range $20°$ to $740°$ are thus

$$z = 196.7°, 376.7°, 556.7°, 736.7°$$

The values of θ in the range $0°$ to $360°$ are found using $\theta = (z - 20°)/2$.

$$\theta = \frac{z - 20°}{2} = 88.35°, 178.35°, 268.35°, 358.35°$$

Exercise 17.5

1. Solve the following trigonometrical equations for $0° \leq \theta \leq 360°$:

   (a) $\sin(2\theta + 50°) = 0.5$
   (b) $\cos(\theta + 110°) = 0.3$

   (c) $\tan\left(\dfrac{\theta}{2}\right) = 1$

   (d) $\sin\left(\dfrac{\theta}{2}\right) = -0.5$

   (e) $\cos(2\theta - 30°) = -0.5$
   (f) $\tan(3\theta - 20°) = 0.25$
   (g) $\sin 2\theta = 2\cos 2\theta$

17.6   Trigonometrical identities

In §17.2 we saw that $\sin(180° - \theta) = \sin\theta$. The left-hand side and the right-hand side have the same value for any value of θ. We call this a **trigonometrical identity**. Other trigonometrical identities studied in §17.2 include $\cos(180° - \theta) = -\cos\theta$ and $\tan(180° - \theta) = -\tan\theta$. Table 17.4 gives many of the most commonly used trigonometrical identities.

Table 17.4

$$\sin(A + B) = \sin A \cos B + \sin B \cos A$$

$$\sin(A - B) = \sin A \cos B - \sin B \cos A$$

$$\sin 2A = 2 \sin A \cos A$$

$$\cos(A + B) = \cos A \cos B - \sin A \sin B$$

$$\cos(A - B) = \cos A \cos B + \sin A \sin B$$

$$\cos 2A = (\cos A)^2 - (\sin A)^2 = \cos^2 A - \sin^2 A$$

$$\tan(A + B) = \frac{\tan A + \tan B}{1 - \tan A \tan B}$$

$$\tan(A - B) = \frac{\tan A - \tan B}{1 + \tan A \tan B}$$

$$\sin A + \sin B = 2 \sin\left(\frac{A + B}{2}\right) \cos\left(\frac{A - B}{2}\right)$$

$$\sin A - \sin B = 2 \sin\left(\frac{A - B}{2}\right) \cos\left(\frac{A + B}{2}\right)$$

$$\cos A + \cos B = 2 \cos\left(\frac{A + B}{2}\right) \cos\left(\frac{A - B}{2}\right)$$

$$\cos A - \cos B = -2 \sin\left(\frac{A - B}{2}\right) \sin\left(\frac{A + B}{2}\right)$$

Note that $(\cos A)^2$ is usually written as $\cos^2 A$ and $(\sin A)^2$ is written as $\sin^2 A$.

There is another identity which deserves special mention:

KEY POINT

$$\sin^2 \theta + \cos^2 \theta = 1$$

This identity shows that $\sin \theta$ and $\cos \theta$ are closely related. Knowing $\sin \theta$ the identity can be used to calculate $\cos \theta$ and vice versa.

Test and assignment exercises 17

1. Solve the following trigonometrical equations for $0° \leq \theta \leq 360°$:

(a) $13 \sin 2\theta = 5$ (b) $\cos\left(\frac{\theta}{3}\right) = -0.7$

(c) $\tan(\theta - 110°) = 1.5$
(d) $\cos 3\theta = 2 \sin 3\theta$

(e) $\sin\left(\frac{\theta}{2} - 30°\right) = -0.65$

2. State all values in the range $0°$ to $360°$ of

(a) $\sin^{-1} 0.9$ (b) $\cos^{-1}(-0.25)$
(c) $\tan^{-1}(-1.3)$ (d) $\sin^{-1}(-0.4)$

(e) $\cos^{-1}\left(\dfrac{1}{3}\right)$ (f) $\tan^{-1} 1.7$

3. Use a graphics calculator or computer graph plotting package to draw the following graphs:

(a) $y = \dfrac{1}{2}\sin 2\theta$ for $0° \leq \theta \leq 360°$

(b) $y = 3\tan\left(\dfrac{\theta}{2}\right)$ for $0° \leq \theta \leq 720°$

(c) $y = \cos(\theta - 90°)$ for $0° \leq \theta \leq 360°$

4. In which quadrant does the angle α lie, given

(a) $\tan \alpha > 0$ and $\cos \alpha < 0$
(b) $\sin \alpha < 0$ and $\tan \alpha < 0$
(c) $\sin \alpha > 0$ and $\cos \alpha < 0$
(d) $\tan \alpha < 0$ and $\cos \alpha < 0$

18 Solution of triangles

Objectives

This chapter

- explains the terms 'scalene', ' isosceles', 'equilateral' and 'right-angled' as applied to triangles
- states Pythagoras's theorem and shows how it can be used in the solution of right-angled triangles
- states the sine rule and the cosine rule and shows how they are used to solve triangles

A triangle is solved when all its angles and the lengths of all its sides are known. Before looking at the various rules used to solve triangles we define the various kinds of triangle.

18.1 Types of triangle

A **scalene triangle** is one in which all the sides are of different length. In a scalene triangle, all the angles are different too. An **isosceles triangle** is one in which two of the sides are of equal length. Figure 18.1 shows an isosceles triangle with $AB = AC$. In an isosceles triangle, there are also two equal angles. In Figure 18.1, $\angle ABC = \angle ACB$. An **equilateral triangle** has three equal sides and three equal angles each 60°. A **right-angled triangle** is one containing a right-angle, that is, an angle of 90°. The side opposite the right-angle is called the **hypotenuse**. Figure 18.2 shows a right-angled triangle with a right-angle at C. The hypotenuse is AB.

In **any** triangle we note the following properties:

Recall from §16.1 that an acute angle is less than 90° and an obtuse angle is greater than 90° and less than 180°.

(i) The sum of the three angles is always 180°.
(ii) The longest side is opposite the largest angle.
(iii) The shortest side is opposite the smallest angle.
(iv) A triangle contains either three acute angles or two acute and one obtuse angle.

Figure 18.1
An isosceles triangle has
two equal sides and two
equal angles

Figure 18.2
△ABC has a right angle
at C; the hypotenuse is
AB

In △ABC, as a shorthand, we often refer to ∠ABC as ∠B, or more simply as B, ∠ACB as C and so on. The sides also have a shorthand notation. In any triangle ABC, AB is always opposite C, AC is always opposite B and BC is always opposite A. Hence we refer to AB as c, AC as b and BC as a. Figure 18.3 illustrates this. We say that A is **included** by AC and AB, B is included by AB and BC and C is included by AC and BC.

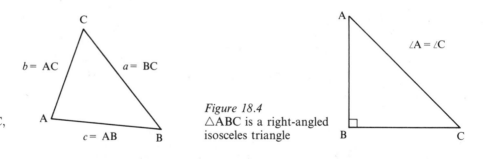

Figure 18.3
△ABC with $a = $ BC,
$b = $ AC, $c = $ AB

Figure 18.4
△ABC is a right-angled
isosceles triangle

Worked example

18.1   State all the angles of a right-angled isosceles triangle.

Solution   Let △ABC be a right-angled isosceles triangle with a right-angle at B. Figure 18.4 illustrates the situation. The angles sum to 180° and $A = C$ because the triangle is isosceles:

$$A + B + C = 180$$
$$A + 90 + C = 180$$
$$A + C = 90$$
$$2A = 90$$
$$A = 45$$

Then $C = 45$ also. The angles are $A = 45°, B = 90°, C = 45°$.

Self-assessment questions 18.1

1. Explain the terms scalene, isosceles and equilateral when applied to triangles.
2. What is the sum of the angles of a triangle?
3. Explain the meaning of the term hypotenuse.
4. Can a triangle containing an obtuse angle possess a hypotenuse?

Exercise 18.1

1. In $\triangle ABC$, $A = 70°$ and $B = 42°$. Calculate C.

2. In $\triangle ABC$, $A = 42°$ and B is twice C. Calculate B and C.

3. An isosceles triangle ABC has $C = 114°$. Calculate A and B.

4. In $\triangle ABC$, $A = 50°$ and $C = 58°$. State (a) the longest side, (b) the shortest side.

5. The smallest angle of an isosceles triangle is $40°$. Calculate the angles of the triangle.

18.2 Pythagoras's theorem

One of the oldest theorems in the study of triangles is due to the ancient Greek mathematician Pythagoras. It applies to right-angled triangles. Consider any right-angled triangle ABC as shown in Figure 18.5.

Figure 18.5
A right-angled triangle

KEY POINT

Pythagoras's theorem states $c^2 = a^2 + b^2$

A **theorem** is an important mathematical result which although not obvious, is possible to prove.

In words we have: the square of the hypotenuse (c^2) equals the sum of the squares of the other two sides ($a^2 + b^2$). Note that as Pythagoras's theorem refers to a hypotenuse, it can be applied only to right-angled triangles.

Worked examples

18.2 In Figure 18.5, AC = 3 cm and BC = 4 cm. Calculate the length of the hypotenuse AB.

Solution We have $a = BC = 4$ and $b = AC = 3$:

$$a^2 = 4^2 = 16 \quad b^2 = 3^2 = 9$$

We apply Pythagoras's theorem:

$$c^2 = a^2 + b^2 = 16 + 9 = 25$$
$$c = 5$$

The hypotenuse, AB, is 5 cm.

18.3 In Figure 18.5, AC = 14 cm and the hypotenuse is 22 cm. Calculate the length of BC.

Solution We have $b = AC = 14$ and $c = AB = 22$. We need to find the length of BC, that is, a. We apply Pythagoras's theorem:

$$c^2 = a^2 + b^2$$
$$22^2 = a^2 + 14^2$$
$$484 = a^2 + 196$$
$$a^2 = 288$$
$$a = \sqrt{288} = 16.97$$

The length of BC is 16.97 cm.

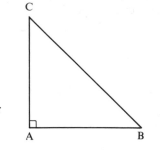

Figure 18.6
Right-angled triangle for
Example 18.4

18.4 ABC is a right-angled isosceles triangle with $A = 90°$. If BC = 12 cm, calculate the lengths of the other sides.

Solution Figure 18.6 illustrates △ABC. The triangle is isosceles and so AC = AB, that is, $b = c$. We are told that the hypotenuse, BC, is 12 cm, that is, $a = 12$. Applying Pythagoras's theorem gives

$$a^2 = b^2 + c^2$$

Substituting $a = 12$ and $b = c$ we get

$$12^2 = b^2 + b^2$$
$$144 = 2b^2$$
$$b^2 = 72$$
$$b = 8.49$$

The sides AC and AB are both 8.49 cm.

Self-assessment questions 18.2

1. State Pythagoras's theorem.
2. Explain what is meant by the hypotenuse of a triangle.

Exercise 18.2

1. $\triangle ABC$ has $C = 90°$. If $AB = 30$ cm and $AC = 17$ cm calculate the length of BC.

2. $\triangle CDE$ has $D = 90°$. Given $CD = 1.2$ m and $DE = 1.7$ m calculate the length of CE.

3. $\triangle ABC$ has a right-angle at B. Given $AC:AB$ is 2:1 calculate AC:BC.

4. $\triangle XYZ$ has $X = 90°$. If $XY = 10$ cm and $YZ = 2XZ$ calculate the lengths of XZ and YZ.

5. $\triangle LMN$ has LN as hypotenuse. Given $LN = 30$ cm and $LM = 26$ cm calculate the length of MN.

18.3   Solution of right-angled triangles

Recall that a triangle is solved when all angles and all lengths have been found. We can use Pythagoras's theorem and the trigonometrical ratios to solve right-angled triangles.

Worked examples

18.5   In $\triangle ABC$, $B = 90°$, $AB = 7$ cm and $BC = 4$ cm. Solve $\triangle ABC$.

Solution   Figure 18.7 shows the information given. We use Pythagoras's theorem to find AC:

$$b^2 = a^2 + c^2$$
$$= 16 + 49 = 65$$
$$b = \sqrt{65} = 8.06$$

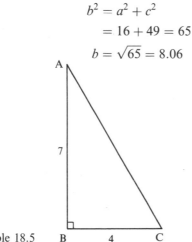

Figure 18.7
$\triangle ABC$ for Example 18.5

We use the trigonometrical ratios to find C:

$$\tan C = \frac{AB}{BC} = \frac{7}{4} = 1.75$$
$$C = \tan^{-1}(1.75) = 60.26°$$

The sum of the angles is 180°:

$$A + B + C = 180$$
$$A + 90 + 60.26 = 180$$
$$A = 29.74$$

The solution to $\triangle ABC$ is

$$
\begin{array}{ll}
A = 29.74° & BC = 4 \text{ cm} \\
B = 90° & AC = 8.06 \text{ cm} \\
C = 60.26° & AB = 7 \text{ cm}
\end{array}
$$

Note that the longest side is opposite the largest angle; the shortest side is opposite the smallest angle.

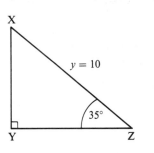

Figure 18.8
$\triangle XYZ$ for Example 18.6

18.6   In $\triangle XYZ$, $Y = 90°$, $Z = 35°$ and $XZ = 10$ cm. Solve $\triangle XYZ$.

Solution   Figure 18.8 illustrates the information given. We find X using the fact that the sum of the angles is 180°:

$$X + Y + Z = 180$$
$$X + 90 + 35 = 180$$
$$X = 55$$

We find the length XY using the sine ratio:

$$\sin Z = \frac{XY}{XZ}$$

$$\sin 35° = \frac{XY}{10}$$

$$XY = 10 \sin 35° = 5.74$$

We find the length YZ using the cosine ratio:

$$\cos Z = \frac{YZ}{XZ}$$

$$\cos 35° = \frac{YZ}{10}$$

$$YZ = 10\cos 35° = 8.19$$

The triangle is now completely solved.

$X = 55°$	YZ = 8.19 cm
$Y = 90°$	XZ = 10 cm
$Z = 35°$	XY = 5.74 cm

Figure 18.9
△ABC for Example 18.7

18.7   In △ABC, $C = 90°$, AB = 20 cm and AC = 14 cm. Solve △ABC.

Solution   Figure 18.9 illustrates the given information. We use Pythagoras's theorem to calculate BC:

$$c^2 = a^2 + b^2$$
$$400 = a^2 + 196$$
$$a^2 = 204$$
$$a = \sqrt{204} = 14.28$$

We use the cosine ratio to calculate A:

$$\cos A = \frac{AC}{AB} = \frac{14}{20} = 0.7$$

$$A = \cos^{-1}(0.7) = 45.57°$$

Since the angles sum to 180° then we have

$$B = 180 - C - A$$
$$= 180 - 90 - 45.57$$
$$= 44.43$$

The solution is

$$A = 45.57° \qquad BC = 14.28 \text{ cm}$$
$$B = 44.43° \qquad AC = 14 \text{ cm}$$
$$C = 90° \qquad AB = 20 \text{ cm}$$

Self-assessment questions 18.3

1. Explain what is meant by 'solving a triangle'.

Exercise 18.3

1. $\triangle ABC$ has $C = 90°$, $A = 37°$ and AC = 36 cm. Solve $\triangle ABC$.

2. $\triangle CDE$ has $E = 90°$, CE = 14 cm and DE = 21 cm. Solve $\triangle CDE$.

3. $\triangle XYZ$ has $X = 90°$, $Y = 26°$ and YZ = 45 mm. Solve $\triangle XYZ$.

4. $\triangle IJK$ has $J = 90°$, IJ = 15 cm and JK = 27 cm. Solve $\triangle IJK$.

5. In $\triangle PQR$, $P = 62°$, $R = 28°$ and PR = 22 cm. Solve $\triangle PQR$.

6. In $\triangle RST$, $R = 70°$, $S = 20°$ and RT = 12 cm. Solve $\triangle RST$.

18.4   The Sine Rule

In §18.3 we saw how to solve right-angled triangles. Many triangles do not contain a right-angle and in such cases we need to use other techniques. One such technique is the **sine rule**.

Consider **any** triangle ABC as shown in Figure 18.3. Recall the notation $a = $ BC, $b = $ AC, $c = $ AB.

KEY POINT

> The sine rule states
> $$\frac{a}{\sin A} = \frac{b}{\sin B} = \frac{c}{\sin C}$$

It must be stressed that the sine rule can be applied to any triangle. The sine rule is used when we are given either (a) two angles and one side or (b) two sides and a non-included angle.

Worked examples

18.8   In △ABC, $A = 30°$, $B = 84°$ and AC = 19 cm. Solve △ABC.

Solution   The information is illustrated in Figure 18.10. We are given two angles and a side and so the sine rule can be used. We can immediately find C since the angles sum to 180°.

$$C = 180° - A - B = 180° - 30° - 84° = 66°$$

We know $A = 30°$, $B = 84°$, $C = 66°$ and $b = AC = 19$. The sine rule states

$$\frac{a}{\sin A} = \frac{b}{\sin B} = \frac{c}{\sin C}$$

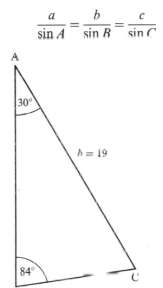

Figure 18.10
△ABC for Example 18.8

Substituting in values gives

$$\frac{a}{\sin 30°} = \frac{19}{\sin 84°} = \frac{c}{\sin 66°}$$

Hence

$$a = \frac{19 \sin 30°}{\sin 84°} = 9.55 \qquad c = \frac{19 \sin 66°}{\sin 84°} = 17.45$$

The solution is

$$\begin{aligned}
A &= 30° & a &= BC = 9.55 \text{ cm} \\
B &= 84° & b &= AC = 19 \text{ cm} \\
C &= 66° & c &= AB = 17.45 \text{ cm}
\end{aligned}$$

18.9   In △ABC, $B = 42°$, AB = 12 cm and AC = 17 cm. Solve △ABC.

Solution   Figure 18.11 illustrates the situation. We know two sides and a non-included angle and so it is appropriate to use the sine rule. We are given $c = AB = 12$, $b = AC = 17$ and $B = 42°$. The sine rule states

$$\frac{a}{\sin A} = \frac{b}{\sin B} = \frac{c}{\sin C}$$

Substituting in the known values gives

$$\frac{a}{\sin A} = \frac{17}{\sin 42°} = \frac{12}{\sin C}$$

Using the equation

$$\frac{17}{\sin 42°} = \frac{12}{\sin C}$$

Figure 18.11
$\triangle ABC$ for Example 18.9

we see that

$$\sin C = \frac{12 \sin 42°}{17} = 0.4723$$

$$C = \sin^{-1}(0.4723) = 28.19° \text{ or } 151.81°$$

As $B = 42°$, then the value 151.81° for C must be rejected as the sum of the angles must be 180°. So $C = 28.19°$. Then

$$A = 180° - B - C = 180° - 42 - 28.19° = 109.81°$$

Using

$$\frac{a}{\sin A} = \frac{17}{\sin 42°}$$

we see that

$$a = \frac{17 \sin A}{\sin 42°}$$

$$= \frac{17 \sin 109.81°}{\sin 42°} = 23.90$$

The solution is

$$A = 109.81° \qquad a = BC = 23.90 \text{ cm}$$
$$B = 42° \qquad\quad b = AC = 17 \text{ cm}$$
$$C = 28.19° \qquad c = AB = 12 \text{ cm}$$

18.10   In $\triangle ABC$, $AB = 23$ cm, $BC = 30$ cm and $C = 40°$. Solve $\triangle ABC$.

Solution   We are told the length of two sides and a non-included angle and so the sine rule can be applied. We have $c = AB = 23$, $a = BC = 30$ and $C = 40°$. The sine rule states

$$\frac{a}{\sin A} = \frac{b}{\sin B} = \frac{c}{\sin C}$$

Substituting in the known values gives

$$\frac{30}{\sin A} = \frac{b}{\sin B} = \frac{23}{\sin 40°}$$

Using the equation

$$\frac{30}{\sin A} = \frac{23}{\sin 40°}$$

we have

$$\sin A = \frac{30 \sin 40°}{23} = 0.8384$$

$$A = \sin^{-1}(0.8384)$$

$$= 56.97° \text{ or } 123.03°$$

There is no reason to reject either of these values and so there are two possible solutions for A. We consider each in turn.

$A = 56.97°$ Here

$$B = 180° - 40° - 56.97° = 83.03°$$

We can now use

$$\frac{b}{\sin B} = \frac{23}{\sin 40°}$$

$$b = \frac{23 \sin B}{\sin 40°} = \frac{23 \sin 83.03°}{\sin 40°}$$

$$= 35.52$$

$A = 123.03°$ In this case

$$B = 180° - 40° - 123.03° = 16.97°$$

We have

$$b = \frac{23 \sin B}{\sin 40°} = \frac{23 \sin 16.97°}{\sin 40°} = 10.44$$

The two solutions are

$$A = 56.97° \qquad a = BC = 30 \text{ cm}$$
$$B = 83.03° \qquad b = AC = 35.52 \text{ cm}$$
$$C = 40° \qquad c = AB = 23 \text{ cm}$$

and

$$A = 123.03° \qquad a = BC = 30 \text{ cm}$$
$$B = 16.97° \qquad b = AC = 10.44 \text{ cm}$$
$$C = 40° \qquad c = AB = 23 \text{ cm}$$

They are illustrated in Figures 18.12(a) and 18.12(b).

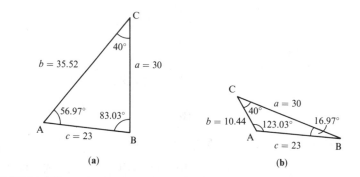

Figure 18.12
Solutions for Example
18.10

Self-assessment questions 18.4

1. State the sine rule and the conditions under which it can be used to solve a triangle.

Exercise 18.4

1. Solve △ABC given

(a) AB = 31, cm AC = 24 cm, $C = 37°$
(b) AB = 19 cm, AC = 24 cm, $C = 37°$
(c) BC = 17 cm, $A = 17°$, $C = 101°$

(d) $A = 53°$, AB = 9.6 cm, BC = 8.9 cm
(e) $A = 62°$, AB = 12.2 cm, BC = 14.5 cm
(f) $B = 36°$, $C = 50°$, AC = 11 cm

18.5   The Cosine Rule

The cosine rule also allows us to solve any triangle. With the usual notation the cosine rule states that:

KEY POINT

$$a^2 = b^2 + c^2 - 2bc \cos A$$
$$b^2 = a^2 + c^2 - 2ac \cos B$$
$$c^2 = a^2 + b^2 - 2ab \cos C$$

This rule is used when we are given either (a) three sides or (b) two sides and the included angle.

Worked examples

18.11   Solve $\triangle ADC$ given AB $-$ 16 cm, AC $=$ 23 cm and BC $=$ 21 cm.

Solution   We are told three sides and so the cosine rule can be used. We have $a = $ BC $= 21$, $b = $ AC $- 23$, $c = $ AB $= 16$. Applying the cosine rule we have

$$a^2 = b^2 + c^2 - 2bc \cos A$$
$$21^2 = 23^2 + 16^2 - 2(23)(16) \cos A$$
$$736 \cos A - 344$$
$$\cos A = \frac{344}{736} = 0.4674$$
$$A - \cos^{-1}(0.4674) - 62.13°$$

We apply the cosine rule again:

$$b^2 = a^2 + c^2 - 2ac \cos B$$
$$23^2 = 21^2 + 16^2 - 2(21)(16) \cos B$$
$$672 \cos B = 168$$
$$\cos B = \frac{168}{672} = 0.25$$
$$B = \cos^{-1}(0.25) = 75.52°$$

Finally

$$C = 180° - A - B = 180° - 62.13° - 75.52° = 42.35°$$

The solution is

$$A = 62.13° \qquad a = \text{BC} = 21 \text{ cm}$$
$$B = 75.52° \qquad b = \text{AC} = 23 \text{ cm}$$
$$C = 42.35° \qquad c = \text{AB} = 16 \text{ cm}$$

18.12   Solve $\triangle ABC$ given $A = 51°$, $AC = 14$ cm and $AB = 24$ cm.

Solution   Figure 18.13 illustrates the situation. We are given two sides and the included angle and so the cosine rule can be used. We have $A = 51°$, $c = AB = 24$ and $b = AC = 14$. We use

$$a^2 = b^2 + c^2 - 2bc \cos A$$

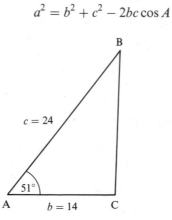

Figure 18.13
$\triangle ABC$ for Example 18.12

Substituting in the values given we have

$$a^2 = 14^2 + 24^2 - 2(14)(24) \cos 51°$$
$$= 196 + 576 - 672 \cos 51°$$
$$= 349.10$$
$$a = 18.68$$

We now use $b^2 = a^2 + c^2 - 2ac \cos B$ to find B:

$$14^2 = (18.68)^2 + 24^2 - 2(18.68)(24) \cos B$$
$$896.64 \cos B = 728.94$$
$$\cos B = \frac{728.94}{896.64} = 0.8130$$
$$B = \cos^{-1}(0.8130) = 35.61°$$

Finally

$$C = 180° - A - B = 180° - 51° - 35.61° = 93.39°$$

The solution is

$A = 51°$	$a = BC = 18.68$ cm
$B = 35.61°$	$b = AC = 14$ cm
$C = 93.39°$	$c = AB = 24$ cm

Self-assessment questions 18.5

1. State the cosine rule and the conditions under which it can be used to solve a triangle.

Exercise 18.5

1. Solve $\triangle ABC$ given

   (a) AC = 119 cm, BC = 86 cm and AB = 53 cm

   (b) AB = 42 cm, AC = 30 cm and $A = 115°$

   (c) BC = 74 cm, AC = 93 cm and $C = 39°$

   (d) AB = 1.9 cm, BC = 3.6 cm and AC = 2.7 cm

   (e) AB = 29 cm, BC = 39 cm and $B = 100°$

Test and assignment exercises 18

1. Solve $\triangle ABC$ given

   (a) $A = 45°, C = 57°$, BC = 19 cm

   (b) AC = 3.9 cm, AB = 4.7 cm, $A = 64°$

   (c) AC = 41 cm, AB = 37 cm, $C = 50°$

   (d) AB = 22 cm, BC = 29 cm, $A = 37°$

   (e) AB = 123 cm, BC = 100 cm, AC = 114 cm

   (f) AB = 46 cm, $A = 33°, B = 76°$

   (g) BC = 64 cm, AC = 54 cm, $B = 42°$

   (h) BC = 30 cm, AC = 69 cm, $C = 97°$

2. $\triangle ABC$ has a right-angle at B. Solve $\triangle ABC$ given

   (a) AB = 9 cm, BC = 14 cm

   (b) BC = 10 cm, AC = 17 cm

   (c) AB = 20 cm, $A = 40°$

   (d) BC = 17 cm, $A = 45°$

   (e) AC = 25 cm, $C = 27°$

   (f) $A = 52°$, AC = 8.6 cm

19 Matrices

Objectives	This chapter
	• explains what is meant by a matrix
	• shows how matrices can be added, subtracted and multiplied
	• explains what is meant by the determinant of a matrix
	• shows how to find the inverse of a matrix
	• shows how matrices can be used to solve simultaneous equations

19.1   What is a matrix?

A **matrix** is a set of numbers arranged in the form of a rectangle and enclosed in curved brackets. The plural of matrix is **matrices**. For example,

$$\begin{pmatrix} 1 & 2 & 3 \\ 3 & 4 & 6 \end{pmatrix} \qquad \begin{pmatrix} 1 \\ 2 \\ -4 \end{pmatrix} \qquad \begin{pmatrix} 1 & 1 & 2 \\ -3 & 4 & 5 \\ \frac{1}{2} & 2 & 1 \end{pmatrix}$$

are all matrices. Each number in a matrix is known as an **element**. To refer to a particular matrix we label it with a capital letter, so that we could write

$$A = \begin{pmatrix} 1 & 2 & 3 \\ 3 & 4 & 6 \end{pmatrix} \qquad B = \begin{pmatrix} 1 \\ 2 \\ -4 \end{pmatrix} \qquad C = \begin{pmatrix} 1 & 1 & 2 \\ -3 & 4 & 5 \\ \frac{1}{2} & 2 & 1 \end{pmatrix}$$

We refer to the **size** of a matrix by giving its number of rows and number of columns, in that order. So matrix A above has two rows and three columns – we say it has size 2×3. Similarly B has size 3×1 and C has size 3×3. Some particular types of matrix occur so frequently that they have been given special names.

A **square** matrix has the same number of rows as columns.

$$\begin{pmatrix} 2 & -7 \\ -1 & 6 \end{pmatrix} \text{ and } \begin{pmatrix} 1 & 2 & -1 \\ 9 & 8 & 5 \\ 6 & -7 & 2 \end{pmatrix}$$

are both square matrices.

A **diagonal** matrix is a square matrix in which all the elements are 0 except those on the diagonal from the top left to the bottom right. This diagonal is called the **leading diagonal**. On the leading diagonal the elements can take any value including zero.

$$\begin{pmatrix} 7 & 0 \\ 0 & 9 \end{pmatrix}, \quad \begin{pmatrix} 9 & 0 \\ 0 & 0 \end{pmatrix} \text{ and } \begin{pmatrix} -1 & 0 & 0 \\ 0 & 8 & 0 \\ 0 & 0 & -17 \end{pmatrix}$$

are all diagonal matrices.

An **identity** matrix is a diagonal matrix, all the diagonal entries of which are equal to 1.

$$\begin{pmatrix} 1 & 0 \\ 0 & 1 \end{pmatrix}, \quad \begin{pmatrix} 1 & 0 & 0 \\ 0 & 1 & 0 \\ 0 & 0 & 1 \end{pmatrix} \text{ and } \begin{pmatrix} 1 & 0 & 0 & 0 \\ 0 & 1 & 0 & 0 \\ 0 & 0 & 1 & 0 \\ 0 & 0 & 0 & 1 \end{pmatrix}$$

are all identity matrices.

Self-assessment questions 19.1

1. Explain what is meant by a matrix.
2. Give an example of a 5×1 matrix and a 1×5 matrix.
3. Explain what is meant by a 'square' matrix.
4. Explain what is meant by the 'leading diagonal' of a square matrix.
5. Explain what is meant by a 'diagonal' matrix.
6. Explain what is meant by an 'identity' matrix.
7. Is it true that all identity matrices must be square?

Exercise 19.1

1. Give the size of each of the following matrices:

$$A = (1 \quad 0 \quad 0 \quad 2) \qquad B = \begin{pmatrix} 1 & 2 \\ 3 & 7 \\ 9 & 8 \\ 9 & -9 \end{pmatrix} \qquad C = (1) \qquad D = \begin{pmatrix} 1 \\ -1 \\ 0 \\ 0 \\ 9 \end{pmatrix}$$

2. Classify each of the following matrices as square, diagonal or identity, as appropriate.

$$A = \begin{pmatrix} 3 & 4 \\ 2 & 9 \end{pmatrix} \qquad B = \begin{pmatrix} 1 & 0 & 0 \\ 0 & 1 & 0 \end{pmatrix} \qquad C = \begin{pmatrix} 1 & 0 & 0 \\ 0 & 1 & 0 \\ 0 & 0 & 1 \end{pmatrix}$$

$$D = \begin{pmatrix} 1 & 5 & 7 \\ 0 & 1 & 0 \end{pmatrix} \qquad E = (1 \ \ 2 \ \ 3 \ \ 4) \qquad F = \begin{pmatrix} 9 & 0 \\ 0 & 8 \end{pmatrix} \qquad G = \begin{pmatrix} 0 & 8 \\ 9 & 0 \end{pmatrix} \qquad H = \begin{pmatrix} 0 & 1 \\ 1 & 0 \end{pmatrix}$$

3. How many elements are there in a matrix whose size is
   (a) 3×1 (b) 1×3 (c) $m \times n$ (d) $n \times n$?

19.2   Addition, subtraction and multiplication of matrices

Matrices can be added, subtracted and multiplied. In this section we shall see how to carry out these operations. However, they can never be divided. Two matrices having the same size can be added or subtracted by simply adding or subtracting the corresponding elements. For example,

$$\begin{pmatrix} 1 & 2 \\ 3 & 4 \end{pmatrix} + \begin{pmatrix} 5 & 2 \\ 1 & 0 \end{pmatrix} = \begin{pmatrix} 1+5 & 2+2 \\ 3+1 & 4+0 \end{pmatrix} = \begin{pmatrix} 6 & 4 \\ 4 & 4 \end{pmatrix}$$

Similarly,

$$\begin{pmatrix} 1 & 2 & 3 & 4 \\ 2 & 1 & 1 & 7 \end{pmatrix} - \begin{pmatrix} 0 & 1 & 3 & 9 \\ 7 & 0 & 0 & 1 \end{pmatrix} = \begin{pmatrix} 1-0 & 2-1 & 3-3 & 4-9 \\ 2-7 & 1-0 & 1-0 & 7-1 \end{pmatrix}$$
$$= \begin{pmatrix} 1 & 1 & 0 & -5 \\ -5 & 1 & 1 & 6 \end{pmatrix}$$

The matrices

$$\begin{pmatrix} 1 & 2 & 9 \\ -1 & \frac{1}{2} & 0 \end{pmatrix} \quad \text{and} \quad \begin{pmatrix} 1 \\ 0 \\ 7 \end{pmatrix}$$

can be neither added to nor subtracted from one another because they have different sizes.

A matrix is multiplied by a number by multiplying each element by that number. For example,

$$4 \begin{pmatrix} 1 & 2 \\ 3 & -9 \end{pmatrix} = \begin{pmatrix} 4 \times 1 & 4 \times 2 \\ 4 \times 3 & 4 \times -9 \end{pmatrix} = \begin{pmatrix} 4 & 8 \\ 12 & -36 \end{pmatrix}$$

and

$$\frac{1}{4} \begin{pmatrix} 16 \\ 8 \end{pmatrix} = \begin{pmatrix} \frac{1}{4} \times 16 \\ \frac{1}{4} \times 8 \end{pmatrix} = \begin{pmatrix} 4 \\ 2 \end{pmatrix}$$

Worked example

19.1   If

$$A = \begin{pmatrix} 5 \\ 1 \\ 9 \end{pmatrix} \quad \text{and} \quad B = \begin{pmatrix} 8 \\ 2 \\ -6 \end{pmatrix} \quad \text{find}$$

(a) $A + B$ (b) $A - B$ (c) $7A$ (d) $-\frac{1}{2}B$ (e) $3A + 2B$

Solution   (a)

$$A + B = \begin{pmatrix} 5 \\ 1 \\ 9 \end{pmatrix} + \begin{pmatrix} 8 \\ 2 \\ -6 \end{pmatrix} = \begin{pmatrix} 13 \\ 3 \\ 3 \end{pmatrix}$$

(b)

$$A - B = \begin{pmatrix} 5 \\ 1 \\ 9 \end{pmatrix} - \begin{pmatrix} 8 \\ 2 \\ -6 \end{pmatrix} = \begin{pmatrix} -3 \\ -1 \\ 15 \end{pmatrix}$$

(c)

$$7A - 7\begin{pmatrix} 5 \\ 1 \\ 9 \end{pmatrix} = \begin{pmatrix} 35 \\ 7 \\ 63 \end{pmatrix}$$

(d)

$$-\frac{1}{2}B = -\frac{1}{2}\begin{pmatrix} 8 \\ 2 \\ -6 \end{pmatrix} = \begin{pmatrix} -4 \\ -1 \\ 3 \end{pmatrix}$$

(e)

$$3A + 2B = 3\begin{pmatrix} 5 \\ 1 \\ 9 \end{pmatrix} + 2\begin{pmatrix} 8 \\ 2 \\ -6 \end{pmatrix} - \begin{pmatrix} 15 \\ 3 \\ 27 \end{pmatrix} + \begin{pmatrix} 16 \\ 4 \\ -12 \end{pmatrix} = \begin{pmatrix} 31 \\ 7 \\ 15 \end{pmatrix}$$

Two matrices can only be multiplied together if the number of columns in the first is the same as the number of rows in the second. The product of two such matrices is a matrix which has the same number of rows as the first matrix and the same number of columns as the second. In symbols, this states that if A has size $p \times q$ and B has size $q \times s$, then AB has size $p \times s$. The way that the multiplication is performed may seem strange at first. You will need to work through several examples to understand how it is done.

Worked examples

19.2   Find AB where

$$A = \begin{pmatrix} 1 & 4 \\ 6 & 3 \end{pmatrix} \quad \text{and} \quad B = \begin{pmatrix} 2 \\ 5 \end{pmatrix}$$

Solution   The size of A is 2×2. The size of B is 2×1. Therefore the number of columns in the first matrix, A, equals the number of rows in the second matrix B. We can therefore find the product AB. The result will be a 2×1 matrix. Let us call it C. The working is as follows.

$$C = \begin{pmatrix} 1 & 4 \\ 6 & 3 \end{pmatrix} \begin{pmatrix} 2 \\ 5 \end{pmatrix} = \begin{pmatrix} 1 \times 2 + 4 \times 5 \\ 6 \times 2 + 3 \times 5 \end{pmatrix} = \begin{pmatrix} 22 \\ 27 \end{pmatrix}$$

The first row of A multiplies the first column of B to give

$$(1 \times 2) + (4 \times 5) = 2 + 20 = 22$$

This is the element in row 1, column 1 of C. The second row of A then multiplies the first column of B to give

$$(6 \times 2) + (3 \times 5) = 12 + 15 = 27$$

This is the element in row 2, column 1 of C.

19.3   Using the same A and B as the previous example, is it possible to find BA?

Solution   Recall that B has size 2×1 and A has size 2×2. When written in the order BA, the number of columns in the first matrix is 1 whilst the number of rows in the second is 2. It is not therefore possible to find BA.

19.4   Find, if possible,

$$\begin{pmatrix} 1 & 4 & 9 \\ 2 & 0 & 1 \end{pmatrix} \begin{pmatrix} 1 & 9 \\ 8 & 7 \\ -7 & 3 \end{pmatrix}$$

Solution   The number of columns in the first matrix is 3 and this is the same as the number of rows in the second. We can therefore perform the multiplication and the answer will have size 2×2.

$$\begin{pmatrix} 1 & 4 & 9 \\ 2 & 0 & 1 \end{pmatrix} \begin{pmatrix} 1 & 9 \\ 8 & 7 \\ -7 & 3 \end{pmatrix} = \begin{pmatrix} 1 \times 1 + 4 \times 8 + 9 \times -7 & 1 \times 9 + 4 \times 7 + 9 \times 3 \\ 2 \times 1 + 0 \times 8 + 1 \times -7 & 2 \times 9 + 0 \times 7 + 1 \times 3 \end{pmatrix}$$

$$= \begin{pmatrix} -30 & 64 \\ -5 & 21 \end{pmatrix}$$

19.5   Find $(1 \quad 5)\begin{pmatrix} 2 \\ 1 \end{pmatrix}$

Solution $(1 \quad 5)\begin{pmatrix} 2 \\ 1 \end{pmatrix} = (1 \times 2 + 5 \times 1) = (7)$

Note that in this example the result is a 1×1 matrix, that is, a single number.

19.6   Find $\begin{pmatrix} 3 & 2 \\ 5 & 3 \end{pmatrix}\begin{pmatrix} x \\ y \end{pmatrix}$.

Solution $\begin{pmatrix} 3 & 2 \\ 5 & 3 \end{pmatrix}\begin{pmatrix} x \\ y \end{pmatrix} = \begin{pmatrix} 3x + 2y \\ 5x + 3y \end{pmatrix}$

19.7   Find IX where I is the 2×2 identity matrix

$$\begin{pmatrix} 1 & 0 \\ 0 & 1 \end{pmatrix} \quad \text{and} \quad X = \begin{pmatrix} a \\ b \end{pmatrix}.$$

Solution $IX = \begin{pmatrix} 1 & 0 \\ 0 & 1 \end{pmatrix}\begin{pmatrix} a \\ b \end{pmatrix} = \begin{pmatrix} 1 \times a + 0 \times b \\ 0 \times a + 1 \times b \end{pmatrix} = \begin{pmatrix} a \\ b \end{pmatrix}$

Note that the effect of multiplying the matrix X by the identity matrix is to leave X unchanged. This is a very important and useful property of identity matrices which we will require later. It should be remembered. This property of identity matrices should remind you of the fact that multiplying a number by 1 leaves the number unchanged: for example, $7 \times 1 = 7$, $1 \times -9 = -9$. An identity matrix plays the same role for matrices as the number 1 does when dealing with ordinary arithmetic.

Self-assessment questions 19.2

1. Under what conditions can two matrices be added or subtracted?
2. Two different types of multiplication have been described. What are these two types and how do they differ?
3. If A has size $p \times q$ and B has size $r \times s$, under what conditions will the product AB exist? If AB does exist what will be its size? Under what conditions will the product BA exist and what will be its size?
4. Suppose A and B are two matrices. The products AB and BA both exist. What can you say about the sizes of A and B?

Exercise 19.2

1. If $M = \begin{pmatrix} 7 & 8 \\ 2 & -1 \end{pmatrix}$ and $N = \begin{pmatrix} 3 & 2 \\ -1 & 4 \end{pmatrix}$ find MN and NM. Comment upon your two answers.

2. If

$$A = \begin{pmatrix} 1 & -7 \\ -1 & 2 \end{pmatrix} \qquad B = \begin{pmatrix} 3 & 7 & 2 \\ -1 & 0 & -10 \end{pmatrix} \qquad C = (1 \quad 0 \quad -1 \quad 1)$$

$$D = \begin{pmatrix} 7 \\ 9 \\ 8 \\ 1 \end{pmatrix} \qquad E = \begin{pmatrix} 1/2 \\ 0 \\ 0 \\ 1 \end{pmatrix}$$

find, if possible,

(a) $3A$ (b) $A + B$ (c) $A + C$ (d) $5D$ (e) BC (f) CB (g) AC (h) CA (i) $5A - 2C$
(j) $D - E$ (k) BA (l) AB (m) AA (that is, A times itself, usually written A^2).

3. I is the 2×2 identity matrix

$$\begin{pmatrix} 1 & 0 \\ 0 & 1 \end{pmatrix}$$

If

$$A = \begin{pmatrix} 3 & 7 \\ -1 & -2 \end{pmatrix} \qquad B = \begin{pmatrix} x \\ y \end{pmatrix} \quad \text{and} \quad C = \begin{pmatrix} 3 & 9 & 5 \\ 4 & 2 & 8 \end{pmatrix}$$

find, where possible,

(a) IA (b) AI (c) BI (d) IB (e) CI (f) IC. Comment upon your answers.

4. If

$$A = \begin{pmatrix} a & b \\ c & d \end{pmatrix} \quad \text{and} \quad B = \begin{pmatrix} e & f \\ g & h \end{pmatrix}$$

find

(a) $A + B$ (b) $A - B$ (c) AB (d) BA

5. Find

$$\begin{pmatrix} -7 & 5 \\ -2 & -1 \end{pmatrix} \begin{pmatrix} x \\ y \end{pmatrix}$$

6. Find

$$\begin{pmatrix} 9 & -5 \\ 3 & 8 \end{pmatrix} \begin{pmatrix} a \\ b \end{pmatrix}$$

19.3   The inverse of a 2 × 2 matrix

Consider the 2 × 2 matrix

$$A = \begin{pmatrix} a & b \\ c & d \end{pmatrix}$$

An important matrix which is related to A is known as the **inverse** of A and is given the symbol A^{-1}. Here, the superscript '−1' should not be read as a power, but is meant purely as a notation for the inverse matrix. A^{-1} can be found from the following formula:

KEY POINT

$$\text{If } A = \begin{pmatrix} a & b \\ c & d \end{pmatrix} \text{ then } A^{-1} = \frac{1}{ad - bc}\begin{pmatrix} d & -b \\ -c & a \end{pmatrix}$$

This formula states that

- the elements on the leading diagonal are interchanged
- the remaining elements change sign
- the resulting matrix is multiplied by $1/(ad - bc)$.

The inverse matrix has the property that.

KEY POINT

$$A A^{-1} = A^{-1} A = I$$

that is, when a 2 × 2 matrix and its inverse are multiplied together the result is the identity matrix. Consider the following example.

Worked example

19.8   Find the inverse of the matrix

$$A = \begin{pmatrix} 6 & 5 \\ 2 & 2 \end{pmatrix}$$

and verify that $A A^{-1} = A^{-1} A = I$.

Solution   Using the formula for the inverse we find

$$A^{-1} = \frac{1}{(6)(2) - (5)(2)} \begin{pmatrix} 2 & -5 \\ -2 & 6 \end{pmatrix} = \frac{1}{2} \begin{pmatrix} 2 & -5 \\ -2 & 6 \end{pmatrix} = \begin{pmatrix} 1 & -\frac{5}{2} \\ -1 & 3 \end{pmatrix}$$

The inverse of $\begin{pmatrix} 6 & 5 \\ 2 & 2 \end{pmatrix}$

is therefore $\begin{pmatrix} 1 & -\frac{5}{2} \\ -1 & 3 \end{pmatrix}$

Evaluating AA^{-1}

we find

$$\begin{pmatrix} 6 & 5 \\ 2 & 2 \end{pmatrix} \begin{pmatrix} 1 & -\frac{5}{2} \\ -1 & 3 \end{pmatrix} = \begin{pmatrix} 6 \times 1 + 5 \times -1 & 6 \times (-\frac{5}{2}) + 5 \times 3 \\ 2 \times 1 + 2 \times -1 & 2 \times (-\frac{5}{2}) + 2 \times 3 \end{pmatrix} = \begin{pmatrix} 1 & 0 \\ 0 & 1 \end{pmatrix}$$

which is the 2×2 identity matrix. Also evaluating $A^{-1}A$ we find

$$\begin{pmatrix} 1 & -\frac{5}{2} \\ -1 & 3 \end{pmatrix} \begin{pmatrix} 6 & 5 \\ 2 & 2 \end{pmatrix} = \begin{pmatrix} 1 \times 6 + (-\frac{5}{2}) \times 2 & 1 \times 5 + (-\frac{5}{2}) \times 2 \\ -1 \times 6 + 3 \times 2 & -1 \times 5 + 3 \times 2 \end{pmatrix} = \begin{pmatrix} 1 & 0 \\ 0 & 1 \end{pmatrix}$$

which is the 2×2 identity matrix. We have shown that
$AA^{-1} = A^{-1}A = I$.

The quantity $ad - bc$ in the formula for the inverse is known as the **determinant** of the matrix A. We often write it as $|A|$, the vertical bars indicating that we mean the determinant of A and not the matrix itself.

KEY POINT

If $A = \begin{pmatrix} a & b \\ c & d \end{pmatrix}$ then its determinant equals

$$|A| = \begin{vmatrix} a & b \\ c & d \end{vmatrix} = ad - bc$$

Worked examples

19.9   Find the value of the determinant

$$\begin{vmatrix} 8 & 7 \\ 9 & 2 \end{vmatrix}$$

Solution   The determinant is given by $(8)(2) - (7)(9) = 16 - 63 = -47$.

19.10 Find the determinant of each of the following matrices:

(a) $C = \begin{pmatrix} 7 & 2 \\ 4 & 9 \end{pmatrix}$ (b) $D = \begin{pmatrix} 4 & -2 \\ 9 & -1 \end{pmatrix}$ (c) $I = \begin{pmatrix} 1 & 0 \\ 0 & 1 \end{pmatrix}$ (d) $E = \begin{pmatrix} 4 & 2 \\ 2 & 1 \end{pmatrix}$

Solution (a) $|C| = (7)(9) - (2)(4) = 63 - 8 = 55$. Note that the determinant is always a single number.

(b) $|D| = (4)(-1) - (-2)(9) = -4 + 18 = 14$.

(c) $|I| = (1)(1) - (0)(0) = 1 - 0 = 1$. It is always true that the determinant of an identity matrix is 1.

(d) $|E| = (4)(1) - (2)(2) = 4 - 4 = 0$. In this example the determinant of the matrix E is zero.

As we have seen in the previous example, on some occasions a matrix may be such that $ad - bc = 0$, that is, its determinant is zero. Such a matrix is called **singular.** When a matrix is singular it cannot have an inverse. This is because it is impossible to evaluate the quantity $1/(ad - bc)$ which appears in the formula for the inverse; the quantity $1/0$ has no meaning in mathematics.

Worked example

19.11 Show that the matrix

$$P = \begin{pmatrix} 6 & 2 \\ -24 & 8 \end{pmatrix}$$

has no inverse.

Solution We first find the determinant of P:

$$|P| = (6)(8) - (-2)(-24) = 48 - 48 = 0.$$

The determinant is zero and so the matrix P is singular. It does not have an inverse.

Self-assessment questions 19.3

1. Explain what is meant by the inverse of a 2 × 2 matrix. Under what conditions will this inverse exist?
2. Explain what is meant by a singular matrix.
3. A diagonal matrix has the elements on its leading diagonal equal to 7 and −9. What is the value of its determinant?

Exercise 19.3

1. Evaluate the following determinants:

(a) $\begin{vmatrix} 9 & 7 \\ -1 & -1 \end{vmatrix}$ (b) $\begin{vmatrix} 8 & 7 \\ 2 & -1 \end{vmatrix}$ (c) $\begin{vmatrix} 0 & 1 \\ -1 & 0 \end{vmatrix}$ (d) $\begin{vmatrix} 8 & 0 \\ 0 & 11 \end{vmatrix}$

2. Which of the following matrices are singular?

(a) $A = \begin{pmatrix} 9 & 5 \\ -2 & 0 \end{pmatrix}$ (b) $B = \begin{pmatrix} 6 & 45 \\ -2 & 15 \end{pmatrix}$ (c) $C = \begin{pmatrix} 6 & 45 \\ 2 & 15 \end{pmatrix}$ (d) $D = \begin{pmatrix} 9 & 5 \\ 0 & 1 \end{pmatrix}$

(e) $E = \begin{pmatrix} a & b \\ a & b \end{pmatrix}$

3. By multiplying the matrices

$\begin{pmatrix} 4 & -9 \\ -3 & 7 \end{pmatrix}$ and $\begin{pmatrix} 7 & 9 \\ 3 & 4 \end{pmatrix}$ together show that each one is the inverse of the other.

4. Find, if it exists, the inverse of each of the following matrices:

(a) $\begin{pmatrix} 2 & 4 \\ 6 & 10 \end{pmatrix}$ (b) $\begin{pmatrix} 1 & 0 \\ 0 & 1 \end{pmatrix}$ (c) $\begin{pmatrix} -1 & 3 \\ 2 & 2 \end{pmatrix}$ (d) $\begin{pmatrix} 8 & 4 \\ 2 & 1 \end{pmatrix}$

19.4   Application of matrices to solving simultaneous equations

Matrices can be used to solve simultaneous equations. Suppose we wish to solve the simultaneous equations

$$3x + 2y = -3$$
$$5x + 3y = -4$$

First of all we rewrite them using matrices as

$$\begin{pmatrix} 3 & 2 \\ 5 & 3 \end{pmatrix}\begin{pmatrix} x \\ y \end{pmatrix} = \begin{pmatrix} -3 \\ -4 \end{pmatrix}$$

Writing

$$A = \begin{pmatrix} 3 & 2 \\ 5 & 3 \end{pmatrix}, \quad X = \begin{pmatrix} x \\ y \end{pmatrix} \quad \text{and} \quad B = \begin{pmatrix} -3 \\ -4 \end{pmatrix}$$

we have $AX = B$. Note that we are trying to find x and y; in other words we must find X. Now, if

$$AX = B$$

then multiplying both sides by A^{-1} gives

$$A^{-1}AX = A^{-1}B$$

But $A^{-1}A = I$, so this becomes

$$IX = A^{-1}B$$

However, multiplying X by the identity matrix leaves X unaltered, so we can write

$$X = A^{-1}B$$

In other words if we multiply B by the inverse of A we will have X as required. Now, given

$$A = \begin{pmatrix} 3 & 2 \\ 5 & 3 \end{pmatrix}$$

then

$$A^{-1} = \frac{1}{(3)(3) - (2)(5)} \begin{pmatrix} 3 & -2 \\ -5 & 3 \end{pmatrix} = \frac{1}{-1} \begin{pmatrix} 3 & -2 \\ 5 & 3 \end{pmatrix} = \begin{pmatrix} -3 & 2 \\ 5 & -3 \end{pmatrix}$$

Finally,

$$X = A^{-1}B - \begin{pmatrix} -3 & 2 \\ 5 & -3 \end{pmatrix}\begin{pmatrix} -3 \\ -4 \end{pmatrix} = \begin{pmatrix} 1 \\ -3 \end{pmatrix}$$

Therefore the solution of the simultaneous equations is $x = 1$ and $y = -3$.

Worked example

19.12   Solve the simultaneous equations

$$x + 2y = 13$$
$$2x - 5y = 8$$

Solution   First of all we rewrite the equations using matrices as

$$\begin{pmatrix} 1 & 2 \\ 2 & -5 \end{pmatrix}\begin{pmatrix} x \\ y \end{pmatrix} = \begin{pmatrix} 13 \\ 8 \end{pmatrix}$$

Writing

$$A = \begin{pmatrix} 1 & 2 \\ 2 & -5 \end{pmatrix}, \quad X = \begin{pmatrix} x \\ y \end{pmatrix} \quad \text{and} \quad B = \begin{pmatrix} 13 \\ 8 \end{pmatrix}$$

we have $AX = B$. We must now find the matrix X. If $AX = B$ then multiplying both sides by A^{-1} gives $A^{-1}AX = A^{-1}B$. But $A^{-1}A = I$, so this becomes $IX = A^{-1}B$. It follows that

$$X = A^{-1}B$$

We must multiply B by the inverse of A. The inverse of A is

$$A^{-1} = \frac{1}{(1)(-5) - (2)(2)} \begin{pmatrix} -5 & -2 \\ -2 & 1 \end{pmatrix} = \frac{1}{-9} \begin{pmatrix} -5 & -2 \\ -2 & 1 \end{pmatrix}$$

Finally,

$$X = A^{-1}B = \frac{1}{-9} \begin{pmatrix} -5 & -2 \\ -2 & 1 \end{pmatrix} \begin{pmatrix} 13 \\ 8 \end{pmatrix} = -\frac{1}{9} \begin{pmatrix} -81 \\ -18 \end{pmatrix} = \begin{pmatrix} 9 \\ 2 \end{pmatrix}$$

Therefore the solution of the simultaneous equations is $x = 9$ and $y = 2$.

Self-assessment questions 19.4

1. Explain how the simultaneous equations $ax + by = c$, $dx + ey = f$, where a, b, c, d, e and f are given numbers and x and y are the unknowns, can be written using matrices.

Exercise 19.4

1. Use matrices to solve the following simultaneous equations:
   (a) $x + y = 17$, $2x - y = 10$ (b) $2x - y = 10$, $x + 3y = -2$
   (c) $x + 2y = 15$, $3x - y = 10$

2. Try to solve the simultaneous equations $2x - 3y = 5$, $6x - 9y = 15$ using the matrix method. What do you find? Can you explain this?

Test and assignment exercises 19

1. Express

   $$\frac{1}{2} \begin{pmatrix} 6 & 6 \\ -18 & 4 \end{pmatrix} + 7 \begin{pmatrix} 8 & 2 \\ 0 & 2 \end{pmatrix} \quad \text{as a single matrix.}$$

2. If

   $$A = \begin{pmatrix} 7 & 7 & 1 \\ 2 & 0 & 9 \end{pmatrix} \quad \text{and } B = \begin{pmatrix} 9 \\ 3 \\ 6 \end{pmatrix}$$

   find, if possible, AB and BA.

3. Find the value of the following determinants:

   (a) $\begin{vmatrix} 9 & -7 \\ 8 & 8 \end{vmatrix}$ (b) $\begin{vmatrix} 9 & 0 \\ 0 & 7 \end{vmatrix}$ (c) $\begin{vmatrix} 0 & 11 \\ -7 & 0 \end{vmatrix}$ (d) $\begin{vmatrix} \alpha & \beta \\ \gamma & \delta \end{vmatrix}$

4. Use matrices to solve the following simultaneous equations:
   (a) $x + 2y = 15$, $x - 2y = -5$ (b) $7x + 8y = 56$, $2x - 2y = 16$
   (c) $2x - 3y = 1$, $x + 2y = -3$

5. Find the inverse of each of the following matrices, if it exists:

   (a) $\begin{pmatrix} -1 & 0 \\ 0 & 1 \end{pmatrix}$ (b) $\begin{pmatrix} 10 & 7 \\ 6 & 4 \end{pmatrix}$ (c) $\begin{pmatrix} \frac{1}{2} & 1 \\ \frac{1}{3} & 7 \end{pmatrix}$

6. Show that each of the following matrices has no inverse:

(a) $\begin{pmatrix} 16 & 8 \\ 4 & 2 \end{pmatrix}$ (b) $\begin{pmatrix} a & b \\ a^2 & ab \end{pmatrix}$

7. If

$$A = \begin{pmatrix} 7 & 5 \\ 2 & -3 \end{pmatrix} \text{ and } B = \begin{pmatrix} -2 & 2 \\ 3 & 7 \end{pmatrix}$$

find AB and BA.

8. Solve the simultaneous equations $2x + 2y = 4$, $7x + 5y = 8$ using the matrix method.

9. Find AB and BA when

$$A - \begin{pmatrix} 1 & 2 & 1 \\ 3 & 4 & 0 \\ 0 & 1 & -1 \end{pmatrix} \quad \text{and} \quad B = \begin{pmatrix} 2 & 4 & 1 \\ 0 & 3 & 1 \\ 0 & 0 & 4 \end{pmatrix}$$

20 Measurement

Objectives

This chapter

- revises common units of length and conversions between them
- revises the concepts of area and volume
- provides a summary of important formulas needed for calculating areas of common shapes and volumes of common solids
- revises common units of mass and conversions between them

20.1 Units of length, area and volume

The standard unit of **length** is the metre, abbreviated to m, although several smaller units are commonly used as well. The following relationships are given for future reference:

KEY POINT

> 1 metre (m) = 100 centimetres (cm)
>   = 1000 millimetres (mm)

Note also that 1000 metres is equal to 1 kilometre (km).

Worked examples

20.1 Convert a length of 6.7 metres into centimetres.

Solution Because 1 m equals 100 cm then 6.7 metres equals $6.7 \times 100 = 670$ cm.

20.2   Convert a length of 24 mm into metres.

Solution   Because 1 m equals 1000 mm, then 1 mm equals $\frac{1}{1000}$ m. Therefore 24 mm equals

$$24 \times \frac{1}{1000} = 0.024 \, m$$

The **area** of a shape is measured by finding how many square units it contains. 1 square metre, written 1 m^2, is the area of a square having sides of length 1 metre. Similarly 1 square centimetre, written 1 cm^2, is the area of a square having sides of length 1 centimetre.

The **volume** of a solid is measured by finding how many cubic units it contains. 1 cubic metre, written 1 m^3, is the volume of a cube with sides of length 1 metre. Similarly, 1 cubic centimetre, written 1 cm^3, is the volume of a cube having sides of length 1 centimetre.

When measuring the volume of liquids, the litre is commonly used. One litre, that is, 1 l, is equal to 1000 cubic centimetres, that is,

$$1 \, l = 1000 \, cm^3$$

Note that 1000 cm^3 is the volume of a cube of side 10 cm, and so it is helpful to remember that such a cube holds 1 litre of liquid. Commonly millilitres are used. 1 millilitre (1 ml) equals 1 cubic centimetre, that is,

$$1 \, ml = 1 \, cm^3$$

KEY POINT

> 1 litre (l) = 1000 millilitres (ml)
>
> 1000 litres (l) = 1 m^3

Worked example

20.3   Convert 256 millilitres into litres.

Solution   Because 1 l = 1000 ml then 1 ml is equal to $\frac{1}{1000}$ l. Therefore

$$256 \, ml = 256 \times \frac{1}{1000} = 0.256 \, l$$

The areas of common shapes and volumes of common solids can be found using formulas. Table 20.1 summarizes a variety of such formulas. The calculation of an area or volume is then simply an exercise in substituting values into a formula.

Table 20.1
Formulae for areas and
volumes of common shapes
and solids

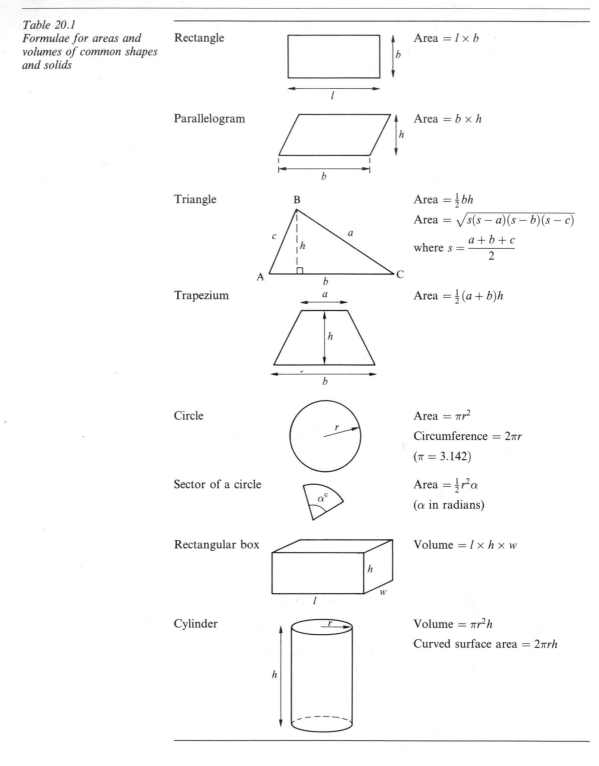

Rectangle		Area $= l \times b$

Parallelogram — Area $= b \times h$

Triangle

Area $= \frac{1}{2}bh$

Area $= \sqrt{s(s-a)(s-b)(s-c)}$

where $s = \dfrac{a+b+c}{2}$

Trapezium — Area $= \frac{1}{2}(a+b)h$

Circle

Area $= \pi r^2$

Circumference $= 2\pi r$

$(\pi = 3.142)$

Sector of a circle

Area $= \frac{1}{2}r^2\alpha$

(α in radians)

Rectangular box — Volume $= l \times h \times w$

Cylinder

Volume $= \pi r^2 h$

Curved surface area $= 2\pi rh$

Table 20.1 (continued)   Cone

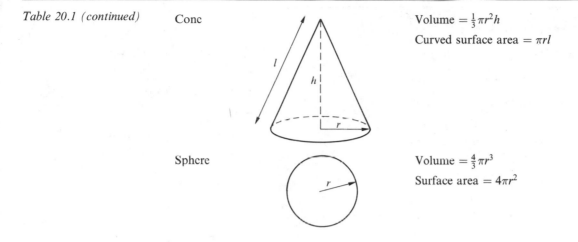

Volume $= \frac{1}{3}\pi r^2 h$

Curved surface area $= \pi r l$

Sphere

Volume $= \frac{4}{3}\pi r^3$

Surface area $= 4\pi r^2$

Worked examples

20.4 A square has sides of length 1 m. Calculate the area in

(a) m² (b) cm² (c) mm²

Solution (a) When working in metres the area is $1 \times 1 = 1\,\text{m}^2$.

(b) When working in centimetres, each side is of length 100 cm. Therefore the area is $100 \times 100 = 10000 = 10^4\,\text{cm}^2$.

(c) When working in millimetres, each side is of length 1000 mm and so the area is

$$1000 \times 1000 = 10^3 \times 10^3 = 10^6\,\text{mm}^2$$

Note that in all three cases the areas are the same and so we can say

$$1\,\text{m}^2 = 10^4\,\text{cm}^2 = 10^6\,\text{mm}^2$$

20.5 A cubic box has sides of length 1 m. Calculate its volume in

(a) m³ (b) cm³ (c) mm³

Solution (a) When working in metres the volume is $1 \times 1 \times 1 = 1\,\text{m}^3$.

(b) When working in centimetres, each side is of length 100 cm. Therefore the volume is $100 \times 100 \times 100 = 1000000 = 10^6\,\text{cm}^3$.

(c) When working in millimetres, each side is of length 1000 mm and so the volume is

$$1000 \times 1000 \times 1000 = 10^3 \times 10^3 \times 10^3 = 10^9\,\text{mm}^3$$

Note that in all three cases the volumes are the same and so we can say

$$1\,\text{m}^3 = 10^6\,\text{cm}^3 = 10^9\,\text{mm}^3$$

Exercise 20.1

1. Convert 16 cm into (a) mm, (b) m.

2. Change 356 mm into (a) cm, (b) m.

3. Change 156 m into km.

4. Find the area of a rectangle whose length is 14 cm and whose width is 7 cm.

5. A room has length 5 m and width 3.5 m. It is required to carpet this room at £4.99 per square metre. Calculate the cost of the carpet assuming there is no wastage due to cutting.

6. Find the area of a parallelogram which has a vertical height of 2 m and a base of length 8 m.

7. A triangle has sides of length 8 cm, 12 cm and 14 cm. Find its area.

8. Find the area of a trapezium whose parallel sides are 14 cm and 8 cm long and whose distance apart is 2 cm.

9. Find the area of a circle whose radius is 18 cm.

10. Find the area of a circle whose diameter is 10 cm.

11. Find the radius of the circle whose area is 26 cm².

12. A circular piece of cardboard of radius 70 cm has a circular piece of radius 25 cm removed from it. Calculate the area of the remaining cardboard.

13. Find the volume of a tin of beans of base radius 5 cm and height 14 cm. What is the tin's capacity in litres?

14. Find the volume of a sphere of radius 8 cm.

15. A sphere has volume 86 cm³. Calculate its radius.

16. Calculate the area of a sector of a circle of angle 45° and radius 10 cm.

17. A garden swimming pool is designed in the form of a cylinder of base radius 3 m. Calculate how many litres of water are required to fill it to a depth of 1 m.

20.2	**Units of mass**

The standard unit of mass is the kilogram, abbreviated to kg. Sometimes smaller units, the gram (g) or milligram (mg), are used. A very sensitive balance may be able to measure in micrograms (μg). The relationship between these units is given as follows:

KEY POINT

> 1 kilogram (kg) = 1000 grams (g) = 1 000 000 milligrams (mg)
> 1 gram (g) = 1000 milligrams (mg) = 1 000 000 micrograms (μg)
> 1 milligram = 1000 micrograms (μg)

Note also that 1000 kilograms equals 1 tonne (t).

Worked examples

20.6   Convert 8.5 kg into grams.

Solution   1 kg is equal to 1000 g and so 8.5 kg must equal $8.5 \times 1000 = 8500$ g.

20.7   Convert 50 g to kilograms.

Solution   Because 1 kg equals 1000 g then 1 g equals $\frac{1}{1000}$ of a kilogram. Consequently,

$$50\,g = 50 \times \frac{1}{1000} = 0.05 \text{ kg}$$

Exercise 20.2

1. Convert 12 kg into (a) grams (b) milligrams.

2. Convert 168 mg into (a) grams (b) kilograms.

3. Convert 0.005 t into kilograms.

4. Convert 875 t into kilograms.

5. Convert 3500 kg into tonnes.

Test and assignment exercises 20

1. Change the following to grams:

   (a) 125 mg (b) 15 kg (c) 0.5 mg

2. Change the following lengths to centimetres:

   (a) 65 mm (b) 0.245 mm (c) 5 m

3. Convert 7500 m and 125.5 m into kilometres.

4. Convert 8 l into millilitres.

5. Convert 56 ml into litres.

6. A household water tank has dimensions 1 m × 0.6 m × 0.8 m. Calculate its capacity in litres assuming it can be filled to the brim.

7. Calculate the area of the sector of a circle of radius 25 cm and angle 60°.

8. A sector of a circle of radius 10 cm has area 186 cm². Calculate the angle of the sector.

9. Calculate the volume of a cylinder whose height is 1 m and whose base radius is 10 cm. (Be careful of the units used.)

10. A sphere has diameter 16 cm. Calculate its volume and surface area.

11. A trapezium has parallel sides of length y_1 and y_2. If these sides are a distance h apart find an expression for the area of the trapezium. Calculate this area when $y_1 = 7$ cm, $y_2 = 10$ cm and $h = 1$ cm.

12. A lorry weighs 7.5 t. Express this in kilograms.

21 Gradients of curves

Objectives

This chapter

- introduces a technique called differentiation for calculating the gradient of a curve at any point
- explains the terms 'maximum' and 'minimum' when applied to functions
- applies the technique of differentiation to locating maximum and minimum values of a function

21.1 The gradient function

Suppose we have a function, $y = f(x)$, and are interested in its slope, or gradient, at several points. For example, Figure 21.1 shows a graph of the function $y = 2x^2 + 3x$. Imagine tracing the graph from the left to the right. At point A the graph is falling rapidly. At point B the graph is falling but less rapidly than at A. Point C lies at the bottom of the dip. At point D the graph is rising. At E it is rising but more quickly than at D. The important point is that the gradient of the curve changes from point

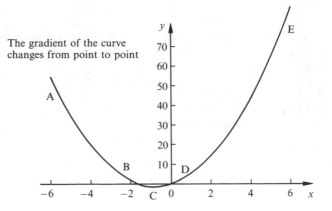

Figure 21.1
Graph of $y = 2x^2 + 3x$

y′ stands for the gradient function of *y = f(x)*. We can also denote the gradient function by *f ′(x)*

to point. The following section describes a mathematical technique for measuring the gradient at different points. We introduce another function called the **gradient function**, which we write as y' and read as '*y* dash'. For almost all the functions you will meet this gradient function can be found by using a formula, applying a rule or checking in a table. Knowing the gradient function we can find the gradient of the curve at any point. The following sections show how to find the gradient function of a number of common functions.

21.2   Gradient function of $y = x^n$

For any function of the form $y = x^n$ the gradient function is found from the following formula:

KEY POINT

$$\text{If } y = x^n \qquad \text{then } y' = nx^{n-1}$$

Worked examples

21.1 Find the gradient function of (a) $y = x^3$, (b) $y = x^4$.

Solution (a) Comparing $y = x^3$ with $y = x^n$ we see that $n = 3$. Then
$y' = 3x^{3-1} = 3x^2$.

(b) Applying the formula with $n = 4$ we find that if $y = x^4$ then
$y' = 4x^{4-1} = 4x^3$.

21.2 Find the gradient function of (a) $y = x^2$, (b) $y = x$.

Solution (a) Applying the formula with $n = 2$ we find that if $y = x^2$ then
$y' = 2x^{2-1} = 2x^1$. Because x^1 is simply x we find that the gradient function is $y' = 2x$.

(b) Applying the formula with $n = 1$ we find that if $y = x^1$ then
$y' = 1x^{1-1} = 1x^0$. Because x^0 is simply 1 we find that the gradient function is $y' = 1$.

Finding the gradient of a graph is now a simple matter. Once the gradient function has been found the gradient at any value of x is found by substituting that value into the gradient function. If, after carrying out this substitution, the result is negative, then the curve is falling. If the result is positive, the curve is rising. The size of the gradient function is a measure of how rapidly this fall or rise is taking place. We write $y'(x = 2)$ or simply $y'(2)$ to denote the value of the gradient function when $x = 2$.

Worked examples

21.3   Find the gradient of $y = x^2$ at the points where

(a) $x = -1$ (b) $x = 0$ (c) $x = 2$ (d) $x = 3$

Solution   From Example 21.2, or from the formula, we know that the gradient function of $y = x^2$ is given by $y' = 2x$.

(a) When $x = -1$ the gradient of the graph is then $y'(-1) = 2(-1) = -2$. The fact that the gradient is negative means that the curve is falling at the point.

(b) When $x = 0$ the gradient is $y'(0) = 2(0) = 0$. The gradient of the curve is zero at this point. This means that the curve is neither falling nor rising.

(c) When $x = 2$ the gradient is $y'(2) = 2(2) = 4$. The fact that the gradient is positive means that the curve is rising.

(d) When $x = 3$ the gradient is $y'(3) = 2(3) = 6$ and so the curve is rising here. Comparing this answer with that of part (c) we conclude that the curve is rising more rapidly at $x = 3$ than at $x = 2$ where the gradient was found to be 4.

   The graph of $y = x^2$ is shown in Figure 21.2. If we compare our results with the graph we see that the curve is indeed falling when $x = -1$ (point A) and is rising when $x = 2$ (point C), and $x = 3$ (point D). At the point where $x = 0$ (point B) the curve is neither rising nor falling.

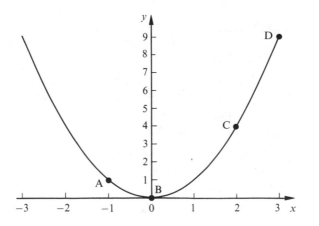

Figure 21.2
Graph of $y = x^2$

21.4   Find the gradient function of $y = x^{-3}$. Hence find the gradient of $y = x^{-3}$ when $x = 4$.

Solution   Using the formula with $n = -3$ we find that if $y = x^{-3}$ then $y' = -3x^{-3-1} = -3x^{-4}$. Because x^{-4} can also be written as $1/x^4$ we could write $y' = -3(1/x^4)$ or $-3/x^4$. When $x = 4$ we find $y'(4) = -3/4^4 = -\frac{3}{256}$. This number is very small and negative, which means that when $x = 4$ the curve is falling, but only slowly.

21.5   Find the gradient function of $y = 1$.

Solution   Before we use the formula to calculate the gradient function let us think about the graph of $y = 1$. Whatever the value of x, this function takes the value 1. Its graph must then be a horizontal line – it neither rises nor falls. We conclude that the gradient function must be zero, that is, $y' = 0$. To obtain the same result using the formula we must rewrite 1 as x^0. Then, using the formula with $n = 0$, we find that if $y = x^0$ then $y' = 0x^{0-1} = 0$.

The previous example illustrates an important result. Any constant function has a gradient function equal to zero because its graph is a horizontal line, and is therefore neither rising nor falling.

The technique introduced in this section has a very long history and as a consequence a wide variety of alternative names for the gradient function have emerged. For example, the gradient function is also called the **first derivative**, or simply the **derivative**. The process of obtaining this is also known as **differentiation.** Being asked to **differentiate** $y = x^5$, say, is equivalent to being asked to find its gradient function, y'. Furthermore, because the gradient function measures how rapidly a graph is changing, it is also referred to as the **rate of change** of y. When you meet these words you should realize that they are simply technical terms associated with finding the gradients of graphs.

Some books use alternative notations for the gradient function. A common alternative is to write y' as dy/dx.

It is possible to find the gradient function of a wide range of functions by simply referring to standard results. Some of these are given in Table 21.1. Note that for technical reasons, when finding the gradient functions of trigonometric functions the angle x must always be measured in radians.

Worked examples

21.6   Use Table 21.1 to find the gradient function y' when y is

(a) $\sin x$ (b) $\sin 2x$ (c) $\cos 3x$ (d) e^x

Table 21.1
The gradient function of some common functions

$y = f(x)$	$y' = f'(x)$	Notes
constant	0	
x	1	
x^2	$2x$	
x^n	nx^{n-1}	
e^x	e^x	
e^{kx}	ke^{kx}	k is a constant
$\sin x$	$\cos x$	
$\cos x$	$-\sin x$	
$\sin kx$	$k\cos kx$	k is a constant
$\cos kx$	$-k\sin kx$	k is a constant
$\ln kx$	$1/x$	k is a constant

Solution (a) Directly from the table we see that if $y = \sin x$ then its gradient function is given by $y' = \cos x$. This result occurs frequently and is worth remembering.

(b) Using the table and taking $k = 2$ we find that if $y = \sin 2x$ then $y' = 2\cos 2x$.

(c) Using the table and taking $k = 3$ we find that if $y = \cos 3x$ then $y' = -3\sin 3x$.

(d) Using the table we see directly that if $y = e^x$ then $y' = e^x$. Note that the exponential function e^x is very special because its gradient function y' is the same as y.

21.7 Find the gradient function of $y = e^{-x}$. Hence find the gradient of the graph of y at the point where $x = 1$.

Solution Noting that $e^{-x} = e^{-1x}$ and using Table 21.1 with $k = -1$ we find that if $y = e^{-x}$ then $y' = -1e^{-x} = -e^{-x}$. Using a calculator to evaluate this when $x = 1$ we find $y'(1) = -e^{-1} = -0.368$.

21.8 Find the gradient function of $y = \sin 4x$ where $x = 0.3$.

Solution Using Table 21.1 with $k = 4$ gives $y' = 4\cos 4x$. Remembering to measure x in radians we evaluate this when $x = 0.3$:

$$y'(0.3) = 4\cos 4(0.3) = 4\cos 1.2 = 1.4494$$

Self-assessment questions 21.2

1. What is the purpose of the gradient function?
2. State the formula for finding the gradient function when $y = x^n$.
3. State three alternative names for the gradient function.
4. State an alternative notation for y'.
5. What is the derivative of any constant? What is the graphical explanation of this answer?
6. When evaluating the derivative of the sine and cosine functions in which units must the angle x be measured – radians or degrees?
7. Which function is equal to its derivative?

Exercise 21.2

1. Find y' when y is given by

   (a) x^0 (b) x^7 (c) x^{-1} (d) x^{-5} (e) x^{13}
   (f) x^5 (g) x^{-2}

2. Find y' when y is given by

   (a) $x^{3/2}$ (b) $x^{5/2}$ (c) $x^{-1/2}$ (d) $x^{1/2}$ (e) $\sqrt{x}$
   (f) $x^{0.2}$

3. Sketch a graph of $y = 8$. State the gradient function, y'.

4. Find the first derivative of each of the following functions:

   (a) $y = x^{16}$ (b) $y = x^{0.5}$ (c) $y = x^{-3.5}$

   (d) $y = \dfrac{1}{x^{1/2}}$ (e) $y = \dfrac{1}{\sqrt{x}}$

5. Find the gradient of the graph of the function $y = x^4$ when

   (a) $x = -4$ (b) $x = -1$ (c) $x = 0$
   (d) $x = 1$ (e) $x = 4$

   Can you infer anything about the shape of the graph from this information?

6. Find the gradient of the graphs of each of the following functions at the points given:

   (a) $y = x^3$ at $x = -3$, at $x = 0$ and at $x = 3$
   (b) $y = x^4$ at $x = -2$ and at $x = 2$

   (c) $y = x^{1/2}$ at $x = 1$ and $x = 2$
   (d) $y = x^{-2}$ at $x = -2$ and $x = 2$

7. Sometimes a function is given in terms of variables other than x. This should pose no problems. Find the gradient functions of

   (a) $y = t^2$ (b) $y = t^{-3}$ (c) $y = t^{1/2}$
   (d) $y = t^{-1.5}$

8. Use Table 21.1 to find the gradient function y' when y is equal to

   (a) $\cos 7x$ (b) $\sin \frac{1}{2}x$ (c) $\cos \frac{x}{2}$
   (d) e^{4x} (e) e^{-3x}

9. Find the derivative of each of the following functions:

   (a) $y = \ln 3x$ (b) $y = \sin \frac{1}{3}x$
   (c) $y = e^{x/2}$

10. Find the gradient of the graph of the function $y = \sin x$ at the points where
    (a) $x = 0$ (b) $x = \pi/2$ (c) $x = \pi$

11. Table 21.1, although written in terms of the variable x, can still be applied when other variables are involved. Find y' when

    (a) $y = t^7$ (b) $y = \sin 3t$ (c) $y = e^{2t}$
    (d) $y = \ln 4t$ (e) $y = \cos \frac{t}{3}$

21.3   Some rules for finding gradient functions

Up to now the formulas we have used for finding gradient functions have applied to single terms. It is possible to apply these much more widely with the addition of three more rules. The first applies to the sum of two functions, the second to their difference and the third to a constant multiple of a function.

KEY POINT

> *Rule 1:*   If $y = f(x) + g(x)$ then $y' = f'(x) + g'(x)$

In words this simply says that to find the gradient function of a sum of two functions we simply find the two gradient functions separately and add these together.

Worked example

21.9   Find the gradient function of $y = x^2 + x^4$.

Solution   The gradient function of x^2 is $2x$. The gradient function of x^4 is $4x^3$. Therefore,

$$\text{if } y = x^2 + x^4 \quad \text{then } y' = 2x + 4x^3$$

The second rule is really an extension of the first, but we state it here for clarity.

KEY POINT

> *Rule 2:*   If $y = f(x) - g(x)$ then $y' = f'(x) - g'(x)$

Worked example

21.10   Find the gradient function of $y = x^5 - x^7$.

Solution   We find the gradient function of each term separately and subtract them. That is, $y' = 5x^4 - 7x^6$.

KEY POINT

> *Rule 3:*   If $y = k f(x)$, where k is a number, then $y' = k f'(x)$

Worked examples

21.11   Find the gradient function of $y = 3x^2$.

Solution   This function is 3 times x^2. The gradient function of x^2 is $2x$. Therefore, using Rule 3, we find that

$$\text{if } y = 3x^2 \qquad \text{then } y' = 3(2x) = 6x$$

The rules can be applied at the same time. Consider the following example.

21.12   Find the derivative of $y = 4x^2 + 3x^{-3}$.

Solution   The derivative of $4x^2$ is $4(2x) = 8x$. The derivative of $3x^{-3}$ is $3(-3x^{-4}) = -9x^{-4}$. Therefore, if $y = 4x^2 + 3x^{-3}$, $y' = 8x - 9x^{-4}$.

21.13   Find the derivative of

(a) $y = 4 \sin t - 3 \cos 2t$ (b) $y = \dfrac{e^{2t}}{3} + 6 + \dfrac{\ln(2t)}{5}$

Solution   (a) We differentiate each quantity in turn using Table 21.1:

$$y' = 4\cos t - 3(-2\sin 2t) = 4\cos t + 6\sin 2t$$

(b) Writing y as $\frac{1}{3}e^{2t} + 6 + \frac{1}{5}\ln(2t)$ we find

$$y' = \frac{2}{3}e^{2t} + 0 + \frac{1}{5}\left(\frac{1}{t}\right) = \frac{2e^{2t}}{3} + \frac{1}{5t}$$

Self-assessment questions 21.3

1. State the rules for finding the derivative of (a) $f(x) + g(x)$ (b) $f(x) - g(x)$ (c) $kf(x)$.

Exercise 21.3

1. Find the gradient function of each of the following:

   (a) $y = x^6 - x^4$ (b) $y = 6x^2$

   (c) $y = 9x^{-2}$ (d) $y = \dfrac{1}{2}x$

   (e) $y = 2x^3 - 3x^2$

2. Find the gradient function of each of the following:

   (a) $y = 2\sin x$ (b) $y = 3\sin 4x$
   (c) $y = 7\cos 9x + 3\sin 4x$
   (d) $y = e^{3x} - 5e^{-2x}$

3. Find the gradient function of $y = 3x^3 - 9x + 2$. Find the value of the gradient function at $x = 1$, $x = 0$ and $x = -1$. At which of these points is the curve steepest?

4. Write down a function which when differentiated gives $2x$. Are there any other functions you can differentiate to give $2x$?

21.4   Higher derivatives

In some applications, and in more advanced work, it is necessary to find the derivative of the gradient function itself. This is termed the **second derivative** and is written y'' and read 'y double dash'. Some books write y'' as d^2y/dx^2.

KEY POINT

$$y'' \text{ or } \frac{d^2y}{dx^2} \text{ is found by differentiating } y'$$

It is a simple matter to find the second derivative by differentiating the first derivative. No new techniques or tables are required.

Worked examples

21.14   Find the first and second derivatives of $y = x^4$.

Solution   From Table 21.1, if $y = x^4$ then $y' = 4x^3$. The second derivative is found by differentiating the first derivative. Therefore

$$\text{if } y' = 4x^3 \quad \text{then } y'' = 4(3x^2) = 12x^2$$

21.15   Find the first and second derivatives of $y = 3x^2 - 7x + 2$.

Solution   Using Table 21.1 we find $y' = 6x - 7$. The derivative of the constant 2 equals 0. Differentiating again we find $y'' = 6$, because the derivative of the constant -7 equals 0.

21.16   If $y = \sin x$, find

(a) $\dfrac{dy}{dx}$ (b) $\dfrac{d^2y}{dx^2}$

Solution   (a) Recall that dy/dx is the first derivative of y. From Table 21.1 this is given by $dy/dx = \cos x$.

(b) d^2y/dx^2 is the second derivative of y. This is found by differentiating the first derivative. From Table 21.1 we find that the derivative of $\cos x$ is $-\sin x$, and so $d^2y/dx^2 = -\sin x$.

Self-assessment questions 21.4

1. Explain how the second derivative of y is found.
2. Give two alternative notations for the second derivative of y.

Exercise 21.4

1. Find the first and second derivatives of the following functions:

   (a) $y = x^2$ (b) $y = 3x^8$ (c) $y = 9x^5$
   (d) $y = 3x^{1/2}$ (e) $y = 3x^4 + 5x^2$
   (f) $y = 9x^3 - 14x + 11$
   (g) $y = x^{1/2} + 4x^{-1/2}$

2. Find the first and second derivatives of

   (a) $y = \sin 3x + \cos 2x$ (b) $y = e^x + e^{-x}$

3. Find the second derivative of $y = \sin 4x$. Show that the sum of $16y$ and y'' is zero, that is $y'' + 16y = 0$.

4. If $y = x^4 - 2x^3 - 36x^2$, find the values of x for which $y'' = 0$.

21.5 Finding maximum and minimum points of a curve

Consider the graph sketched in Figure 21.3. There are a number of important points marked on this graph, all of which have one thing in common. At each point A, B, C and D, the gradient of the graph is zero. If you were travelling from the left to the right then at these points the graph would appear to be flat. Points where the gradient is zero are known as **stationary points**. To the left of point A, the curve is rising; to its right the curve is falling. A point such as A is called a **maximum turning point** or simply a **maximum**. To the left of point C the curve is falling; to its right the curve is rising. A point such as C is called a **minimum turning point**, or simply a **minimum**. Points B and D are known as **points of inflexion**. At these points the slope of the curve is momentarily zero but then the curve continues rising or falling as before.

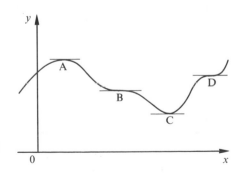

Figure 21.3
The curve has a maximum at A, a minimum at C and points of inflexion at B and D

Because at all these points the gradient is zero, they can be located by looking for values of x which make the gradient function zero.

KEY POINT

Stationary points are located by setting the gradient function equal to zero, that is, $y' = 0$.

Worked example

21.17   Find the stationary points of $y = 3x^2 - 6x + 8$.

Solution   We first determine the gradient function y' by differentiating y. This is found to be $y' = 6x - 6$. Stationary points occur when the gradient is zero, that is, when $y' = 6x - 6 = 0$. From this

$$6x - 6 = 0$$
$$6x = 6$$
$$x = 1$$

When $x = 1$ the gradient is zero, and therefore $x = 1$ is a stationary point. At this stage we cannot tell whether to expect a maximum, minimum or point of inflexion; all we know is that one of these occurs when $x = 1$. However, a sketch of the graph of $y = 3x^2 - 6x + 8$ is shown in Figure 21.4 which reveals that the point is a minimum.

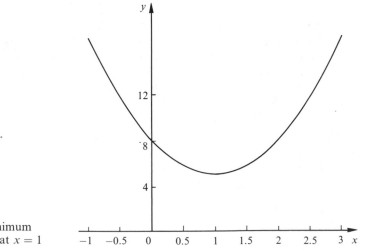

Figure 21.4
There is a minimum turning point at $x = 1$

There are a number of ways to determine the nature of a stationary point once its location has been found. One way, as we have seen, is to sketch a graph of the function. Two alternative methods are now described.

Finding the nature of a stationary point by looking at the gradient on either side

At a stationary point we know that the gradient is zero. If this stationary point is a minimum, then we see from Figure 21.5 that, as we move from left to right, the gradient changes from negative to zero to positive. Therefore a little to the left of a minimum the gradient function will be negative; a little to the right it will be positive. Conversely, if the stationary point is a maximum, as we move from left to right, the gradient

Figure 21.5
The sign of the gradient function close to a stationary point

changes from positive to zero to negative. A little to the left of a maximum the gradient function will be positive; a little to the right it will be negative.

The behaviour of the gradient function close to a point of inflexion is also shown. The sign of the gradient function is the same on both sides of a point of inflexion. With this information we can determine the nature of a stationary point without first plotting a graph.

Worked examples

21.18 Find the location and nature of the stationary points of
$y = 2x^3 - 6x^2 - 18x$.

Solution In order to find the location of the stationary points we must calculate the gradient function. This is $y' = 6x^2 - 12x - 18$. At a stationary point $y' = 0$ and so the equation which must be solved is $6x^2 - 12x - 18 = 0$. This quadratic equation is solved as follows:

$$6x^2 - 12x - 18 = 0$$

factorizing:
$$6(x^2 - 2x - 3) = 0$$
$$6(x + 1)(x - 3) = 0$$
$$\text{so that } x = -1 \text{ and } 3$$

The stationary points are located at $x = -1$ and $x = 3$. We now find their nature.

When $x = -1$ A little to the left of $x = -1$, say at $x = -2$, we calculate the gradient of the graph using the gradient function. That is, $y'(-2) = 6(-2)^2 - 12(-2) - 18 = 24 + 24 - 18 = 30$. Therefore the graph is rising when $x = -2$. A little to the right of $x = -1$, say at $x = 0$, we calculate the gradient of the graph using the gradient function. That is, $y'(0) = 6(0)^2 - 12(0) - 18 = -18$. Therefore the graph is falling when $x = 0$. From this information we conclude that the turning point at $x = -1$ must be a maximum.

When $x = 3$ A little to the left of $x = 3$, say at $x = 2$, we calculate the gradient of the graph using the gradient function. That is, $y'(2) = 6(2)^2 - 12(2) - 18 = 24 - 24 - 18 = -18$. Therefore the graph is falling when $x = 2$. A little to the right of $x = 3$, say at $x = 4$, we calculate the gradient of the graph using the gradient function. That is, $y'(4) = 6(4)^2 - 12(4) - 18 = 96 - 48 - 18 = 30$. Therefore the graph is rising when $x = 4$. From this information we conclude that the turning point at $x = 3$ must be a minimum.

To show this behaviour a graph of $y = 2x^3 - 6x^2 - 18x$ is shown in Figure 21.6 where these turning points can be clearly seen.

21.19 Find the location and nature of the stationary points of $y = x^3 - 3x^2 + 3x - 1$.

Solution First we find the gradient function: $y' = 3x^2 - 6x + 3$. At a stationary point $y' = 0$ and so

$$3x^2 - 6x + 3 = 0$$
$$3(x^2 - 2x + 1) = 0$$
$$3(x - 1)(x - 1) = 0$$
$$\text{and so } x = 1$$

We conclude that there is only one stationary point, and this is at $x = 1$. To determine its nature we look at the gradient function on either side of this point. At $x = 0$, say, $y'(0) = 3$. At $x = 2$, say,

Figure 21.6
Graph of
$y = 2x^3 - 6x^2 - 18x$

$y'(2) = 3(2^2) - 6(2) + 3 - 3$. The gradient function changes from positive to zero to positive as we move through the stationary point. We conclude that the stationary point is a point of inflexion. For completeness a graph is sketched in Figure 21.7.

Figure 21.7
Graph of
$y = x^3 - 3x^2 + 3x - 1$
showing the point of
inflexion

The second derivative test for maximum and minimum points

A simple test exists which uses the second derivative for determining the nature of a stationary point once it has been located.

KEY POINT

If y'' is positive at the stationary point, the point is a minimum.
If y'' is negative at the stationary point, the point is a maximum.
If y'' is equal to zero, this test does not tell us anything and the previous method should be used.

Worked example

21.20   Locate the stationary points of

$$y = \frac{x^3}{3} + \frac{x^2}{2} - 12x + 5$$

and determine their nature using the second derivative test.

Solution   We first find the gradient function: $y' = x^2 + x - 12$. Setting this equal to zero to locate the stationary points we find

$$x^2 + x - 12 = 0$$
$$(x + 4)(x - 3) = 0$$
$$x = -4 \text{ and } 3$$

There are two stationary points: at $x = -4$ and at $x = 3$. To apply the second derivative test we need to find y''. This is found by differentiating y' to give $y'' = 2x + 1$. We now evaluate this at each stationary point in turn.

$x = -4$ When $x = -4$, $y''(-4) = 2(-4) + 1 = -7$. This is negative and we conclude from the second derivative test that the point is a maximum point.

$x = 3$ When $x = 3$, $y''(3) = 2(3) + 1 = 7$. This is positive and we conclude from the second derivative test that the point is a minimum point. Remember, if in doubt, these results could be confirmed by sketching a graph of the function.

Self-assessment questions 21.5

1. State the three types of stationary point.
2. State the second derivative test for determining the nature of a stationary point.
3. Under what condition will the second derivative test fail?

Exercise 21.5

1. Determine the location and nature of any stationary point of the following functions:

   (a) $y = x^2 + 1$ (b) $y = -x^2$
   (c) $y = 2x^3 + 9x^2$ (d) $y = -2x^3 + 27x^2$
   (e) $y = x^3 - 3x^2 + 3x + 1$

Test and assignment exercises 21

1. Find y' when y equals

   (a) $7x$ (b) x^{15} (c) $x^{3/2}$ (d) $3x^{-1}$ (e) $5x^4$
   (f) $16x^{-3}$

2. Find y' when y equals

   (a) $\dfrac{x+1}{x}$ (b) $3x^2 - 7x + 14$ (c) $\sin 3x$

   (d) $e^{5x} + e^{4x}$ (e) $3e^{3x}$

3. Find the first and second derivatives of

   (a) $f(x) = e^{-3x}$ (b) $f(x) = \ln 2x$

   (c) $f(x) = x + \dfrac{x^2}{2} + \dfrac{x^3}{3}$

   (d) $f(x) = -15$ (e) $f(x) = 3x$

4. Find y' when

   (a) $y = 3t - 7$ (b) $y = 16t^2 + 7t + 9$

5. Find the first derivative of each of the following functions:

   (a) $f(x) = x(x+1)$ (b) $f(x) = x(x-2)$
   (c) $f(x) = x^2(x+1)$

   (d) $f(x) = (x-2)(x+3)$
   (e) $f(t) = 3t^2 + 7t + 11$
   (f) $f(t) = 7(t-1)(t+1)$

6. Find the first and second derivatives of each of the following:

   (a) $2x^2$ (b) $4x^3$ (c) $8x^7 - 5x^2$ (d) $\dfrac{1}{x}$

7. Find the location and nature of the maxima and minima of
   $y = 2x^3 + 15x^2 - 84x$.

8. Find a function which when differentiated gives $\frac{1}{2}x^3$.

9. Find the gradient of y at the specified points:

   (a) $y = 3x + x^2$ at $x = -1$.

   (b) $y = 4\cos 3x + \dfrac{\sin 4x}{2}$ at $x = 0.5$.

   (c) $y = 3e^{2x} - 2e^{-3x}$ at $x = 0.1$.

22 Integration and areas under curves

Objectives	This chapter
	• introduces the reverse process of differentiation which is called 'integration'
	• explains what is meant by an 'indefinite integral' and a 'definite integral'
	• shows how integration can be used to find the area under a curve

22.1 Introduction

In some applications it is necessary to find the area under a curve and above the x axis between two values of x, say a and b. Such an area is shown shaded in Figure 22.1. In this chapter we shall show how this area can be found using a technique known as **integration**. However, before we do this it is necessary to consider two types of integration: **indefinite** and **definite** integration.

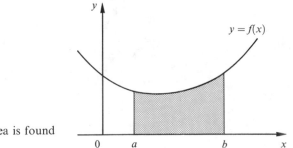

Figure 22.1
The shaded area is found by integration

22.2	**Indefinite integration: the reverse of differentiation**

Suppose we ask 'what function can we differentiate to give $2x$?' In fact, there are lots of correct answers to this question. We can use our knowledge of differentiation to find these. The function $y = x^2$ has derivative $dy/dx = 2x$, so clearly x^2 is one answer. Furthermore, $y = x^2 + 13$ also has derivative $2x$ as do $y = x^2 + 5$ and $y = x^2 - 7$, because differentiation of any constant equals zero. Indeed any function of the form $y = x^2 + c$ where c is a constant has derivative equal to $2x$. The answer to the original question 'what function can we differentiate to give $2x$?' is then any function of the form $y = x^2 + c$. Here we have started with the derivative of a function and then found the function itself. This is differentiation in reverse. We call this process **integration** or more precisely **indefinite integration**. We say that $2x$ has been **integrated** with respect to x to give $x^2 + c$. The quantity $x^2 + c$ is called the **indefinite integral** of $2x$. Both differentiation and indefinite integration are shown schematically in Figure 22.2.

Figure 22.2
The indefinite integral of
$2x$ is $x^2 + c$

We now introduce a notation used to state this compactly. We write

$$\int 2x \, dx = x^2 + c$$

The $\int$ sign indicates that the reverse process to differentiation is to be performed. Following the function to be integrated is always a term of the form dx, which indicates the independent variable, in this case x. When finding indefinite integrals there will always be a constant known as the **constant of integration**. Every time we integrate, the constant of integration must be added.

As a further illustration, if $y = \sin x$ then we know from a table of derivatives that $y' = \cos x$. We can write this as

$$\int \cos x \, dx = \sin x + c$$

To find integrals it is often possible to use a table of derivatives such as that on page 208, and to read the table from right to left. However, it is more helpful to have a table of integrals such as that given in Table 22.1. As with the table of derivatives given in Chapter 21, when dealing with

Table 22.1
Table of indefinite
integrals

$f(x)$	$\int f(x)\,dx$, all '+c'	Notes
$2x$	x^2	
x	$\frac{1}{2}x^2$	
k, constant	kx	
x^n	$\frac{1}{n+1}x^{n+1}$	$n \neq -1$
$\frac{1}{x}$	$\ln x$	
e^x	e^x	
e^{kx}	$\frac{e^{kx}}{k}$	k is a constant
$\sin x$	$-\cos x$	
$\cos x$	$\sin x$	
$\sin kx$	$-\dfrac{\cos kx}{k}$	k is a constant
$\cos kx$	$\dfrac{\sin kx}{k}$	k is a constant

trigonometric functions the angle x must always be measured in radians. The rule

$$\int x^n dx = \frac{1}{n+1}x^{n+1} + c$$

in the table applies for all values of n except $n = -1$.

Worked examples

22.1 Use the table to find the following indefinite integrals:

(a) $\displaystyle\int x^4\,dx$ (b) $\displaystyle\int x^{-2}\,dx$

Solution (a) From the table we note that

$$\int x^n\,dx = \frac{1}{n+1}x^{n+1} + c$$

Therefore, using this formula with $n = 4$, we find

$$\int x^4\,dx = \frac{1}{4+1}x^{4+1} + c = \frac{1}{5}x^5 + c$$

(b) Similarly, using the formula with $n = -2$ we find

$$\int x^{-2}dx = \frac{1}{-2+1}x^{-2+1} + c = \frac{1}{-1}x^{-1} + c = -x^{-1} + c$$

Note that because integration is the reverse of differentiation, in both cases the answer can be checked by differentiating.

22.2 Use Table 22.1 to find

(a) $\int e^x\,dx$ (b) $\int x^{-1}dx$.

Solution (a) Directly from the table we see that $\int e^x dx = e^x + c$.

(b) Noting that $x^{-1} = 1/x$, Table 22.1 gives $\int x^{-1}dx = \ln x + c$. It is important to realize that the formula for integrating x^n cannot be used here; the formula specifically excludes use of the value $n = -1$.

22.3 Use Table 22.1 to find the following indefinite integrals:

(a) $\int e^{2x}dx$ (b) $\int \cos 3x\,dx$

Solution (a) Using Table 22.1 we note that

$$\int e^{kx}dx = \frac{e^{kx}}{k} + c$$

Using this result with $k = 2$ we find

$$\int e^{2x}dx = \frac{e^{2x}}{2} + c$$

(b) Using Table 22.1 we note that

$$\int \cos kx\,dx = \frac{\sin kx}{k} + c$$

Using this result with $k = 3$ we find

$$\int \cos 3x\,dx = \frac{\sin 3x}{3} + c$$

22.4 Find $\int \sqrt{x}\,dx$.

Solution Recall that $\sqrt{x}$ is the same as $x^{1/2}$. Therefore

$$\int \sqrt{x}\,dx = \int x^{1/2}dx = \frac{x^{3/2}}{3/2} + c = \frac{2}{3}x^{3/2} + c$$

Self-assessment questions 22.2

1. Explain what you understand by an 'indefinite integral'.
2. Why is it always necessary to add a constant of integration when finding an indefinite integral?

Exercise 22.2

1. Use Table 22.1 to find the following indefinite integrals:

(a) $\int x \, dx$ (b) $\int 2x \, dx$ (c) $\int x^5 \, dx$

(d) $\int x^7 \, dx$ (e) $\int x^{-5} \, dx$ (f) $\int x^{-1/2} \, dx$

(g) $\int x^0 \, dx$ (h) $\int x^{-1/4} \, dx$

2. Find:

(a) $\int 9 \, dx$ (b) $\int \frac{1}{2} \, dx$ (c) $\int -7 \, dx$

(d) $\int 0.5 \, dx$

3. Use Table 22.1 to find the following integrals:

(a) $\int \sin 5x \, dx$ (b) $\int \sin \frac{1}{2}x \, dx$

(c) $\int \cos \frac{x}{2} \, dx$ (d) $\int e^{-3x} \, dx$

(e) $\int \sin 0.25x \, dx$ (f) $\int e^{-0.5x} \, dx$

(g) $\int e^{x/2} \, dx$

22.3   Some rules for finding other indefinite integrals

Up to now the formulas we have used for finding indefinite integrals have applied to single terms. More complicated functions can be integrated if we make use of three simple rules. The first applies to the sum of two functions, the second to their difference and the third to a constant multiple of a function.

KEY POINT

Rule 1:

$$\int f(x) + g(x) \, dx = \int f(x) \, dx + \int g(x) \, dx$$

In words, this simply says that to find the integral of a sum of two functions we simply find the integrals of the two separate parts and add these together.

Worked example

22.5   Find the integral $\int x^2 + x^3 \, dx$.

Solution   From Table 22.1 the integral of x^2 is $x^3/3 + c_1$. The integral of x^3 is $x^4/4 + c_2$. Note that we have called the constants of integration c_1 and c_2. The integral of $x^2 + x^3$ is simply the sum of these separate integrals, that is,

$$\int x^2 + x^3 \, dx = \frac{x^3}{3} + \frac{x^4}{4} + c_1 + c_2$$

The two constants can be combined into a single constant c, say, so that we would usually write

$$\int x^2 + x^3 \, dx = \frac{x^3}{3} + \frac{x^4}{4} + c$$

The second rule is really an extension of the first, but we state it here for clarity.

KEY POINT

> *Rule 2:*
>
> $$\int f(x) - g(x) \, dx = \int f(x) \, dx - \int g(x) \, dx$$

In words, this says that to find the integral of the difference of two functions we simply find the integrals of the two separate parts and find their difference.

Worked examples

22.6   Find the integral $\int x^3 - x^6 \, dx$.

Solution   From Table 22.1 the integral of x^3 is $x^4/4 + c_1$. The integral of x^6 is $x^7/7 + c_2$. Note once again that we have called the constants of integration c_1 and c_2. The integral of $x^3 - x^6$ is simply the difference of these separate integrals, that is,

$$\int x^3 - x^6 \, dx = \frac{x^4}{4} - \frac{x^7}{7} + c_1 - c_2$$

As before the two constants can be combined into a single constant c, say, so that we would usually write

$$\int x^3 - x^6 \, dx = \frac{x^4}{4} - \frac{x^7}{7} + c$$

22.7   Find $\int \sin 2x - \cos 2x \, dx$.

Solution   Using the second rule we can find the integral of each term separately. From Table 22.1 $\int \sin 2x \, dx = -(\cos 2x)/2 + c_1$ and $\int \cos 2x \, dx = (\sin 2x)/2 + c_2$. Therefore

$$\int \sin 2x - \cos 2x \, dx = \left(-\frac{\cos 2x}{2} + c_1 \right) - \left(\frac{\sin 2x}{2} + c_2 \right)$$

$$= -\frac{\cos 2x}{2} - \frac{\sin 2x}{2} + c$$

where we have combined the two constants into a single constant c.

KEY POINT

Rule 3: If k is a constant, then

$$\int k \times f(x) \, dx = k \times \int f(x) \, dx$$

In words, the integral of k times a function is simply k times the integral of the function. This means that a constant factor can be taken through the integral sign and written at the front, as in the following example.

Worked example

22.8   Find $\int 3x^4 \, dx$.

Solution   Using the third rule we can write

$$\int 3x^4 \, dx = 3 \times \int x^4 \, dx$$

taking the constant factor out. Now $\int x^4 \, dx = x^5/5 + c$ so that

$$\int 3x^4 \, dx = 3 \times \int x^4 \, dx = 3 \times \left(\frac{x^5}{5} + c \right)$$

$$= \frac{3x^5}{5} + 3c$$

It is common practice to write the constant $3c$ as simply another constant K, say, to give $\int 3x^4 \, dx = 3x^5/5 + K$.

Self-assessment questions 22.3

1. State the rules for finding the indefinite integrals $\int f(x) + g(x) \, dx$, $\int f(x) - g(x) \, dx$ and $\int k \times f(x) \, dx$ where k is a constant.

Exercise 22.3

1. Using Table 22.1 and the rules of integration given in this section, find

(a) $\int x^2 + x\,dx$ (b) $\int x^2 + x + 1\,dx$

(c) $\int 4x^6\,dx$ (d) $\int 5x + 7\,dx$

(e) $\int \frac{1}{x}\,dx$ (f) $\int \frac{1}{x^2}\,dx$ (g) $\int \frac{3}{x} + \frac{7}{x^2}\,dx$

(h) $\int x^{1/2}\,dx$ (i) $\int 4x^{-1/2}\,dx$ (j) $\int \frac{1}{\sqrt{x}}\,dx$

(k) $\int 3x^3 + 7x^2\,dx$ (l) $\int 2x^4 - 3x^2\,dx$

(m) $\int x^5 - 7\,dx$

2. Find the following integrals:

(a) $\int 3e^x\,dx$ (b) $\int \frac{1}{2}e^x\,dx$

(c) $\int 3e^x - 2e^{-x}\,dx$ (d) $\int 5e^{2x}\,dx$

(e) $\int \sin x - \cos 3x\,dx$

3. Integrate each of the following functions with respect to x:

(a) $3x^2 - 2e^x$ (b) $\frac{x}{2} + 1$

(c) $2(\sin 3x - 4\cos 3x)$

(d) $3\sin 2x + 5\cos 3x$ (e) $\frac{5}{x}$

4. Find $\int \frac{5}{t} - t^2\,dt$

22.4   Definite integrals

There is a second form of integral known as a definite integral. This is written as

$$\int_a^b f(x)\,dx$$

and is called the **definite integral** of $f(x)$ between the **limits** $x = a$ and $x = b$. This definite integral looks just like the indefinite integral $\int f(x)\,dx$ except values a and b are placed below and above the integral sign. The value $x = a$ is called the **lower limit** and $x = b$ is called the **upper limit**. Unlike the indefinite integral, which is a function depending upon x, the definite integral has a specific numerical value. To find this value we first find the indefinite integral of $f(x)$, and then calculate

its value at the upper limit − its value at the lower limit

The resulting number is the value of the definite integral. To see how this value is obtained consider the following examples.

Worked examples

22.9   Evaluate the definite integral $\int_0^1 x^3\,dx$.

Solution   We first find the indefinite integral $\int x^3\,dx$. This is

$$\int x^3 \, dx = \frac{x^4}{4} + c$$

This is evaluated at the upper limit where $x = 1$, giving $\frac{1}{4} + c$, and at the lower limit where $x = 0$, giving $0 + c$, or simply c. The difference is found:

value at the upper limit $-$ value at the lower limit

$$= \left(\frac{1}{4} + c\right) - (c) = \frac{1}{4}$$

We note that the constant c has cancelled out; this will always happen and so for definite integrals it is usually omitted altogether. The calculation is written more compactly as

When evaluating definite integrals, the constant of integration may be omitted

$$\int_0^1 x^3 \, dx = \left[\frac{x^4}{4}\right]_0^1 \quad \text{the indefinite integral of } x^3 \text{ is written in square brackets}$$

$$= \frac{1}{4} - 0$$

$$= \frac{1}{4}$$

Note that the definite integral $\int_0^1 x^3 \, dx$ has a definite numerical value, $\frac{1}{4}$.

22.10   Find the definite integral $\int_{-1}^1 3x^2 + 7 \, dx$.

Solution

$$\int_{-1}^1 3x^2 + 7 \, dx = [x^3 + 7x]_{-1}^1 \qquad \text{(the indefinite integral of } 3x^2 + 7 \text{ is written in square brackets)}$$

$$= (\text{value of } x^3 + 7x \text{ when } x = 1)$$
$$\quad - (\text{value of } x^3 + 7x \text{ when } x = -1)$$
$$= (1^3 + 7) - \{(-1)^3 - 7\}$$
$$= (8) - (-8)$$
$$= 16$$

22.11   Evaluate $\int_0^1 e^x + 1 \, dx$.

Solution

$$\int_0^1 e^x + 1 \, dx = [e^x + x]_0^1$$

$$= (e^1 + 1) - (e^0 + 0)$$
$$= e^1 + 1 - 1$$
$$= e^1$$
$$= 2.718$$

22.12   Find $\int_{\pi/4}^{\pi/2} \cos 2x \, dx$.

Solution   Note that the calculation must be carried out using radians and not degrees.

$$\int_{\pi/4}^{\pi/2} \cos 2x \, dx = \left[\frac{\sin 2x}{2}\right]_{\pi/4}^{\pi/2}$$

$$= \left(\frac{\sin \pi}{2}\right) - \left(\frac{\sin \pi/2}{2}\right)$$

$$= 0 - \frac{1}{2}$$

$$= -\frac{1}{2}$$

Self-assessment questions 22.4

1. Explain the important way in which a definite integral differs from an indefinite integral.

Exercise 22.4

1. Evaluate the following definite integrals:

(a) $\int_{1}^{1} 7x \, dx$ (b) $\int_{0}^{3} 7x \, dx$ (c) $\int_{-1}^{1} x^2 \, dx$

(d) $\int_{-1}^{1} (-x^2) \, dx$ (e) $\int_{0}^{2} x^2 \, dx$

(f) $\int_{0}^{3} x^2 + 4x \, dx$ (g) $\int_{0}^{\pi} \cos x \, dx$

(h) $\int_{1}^{2} \frac{1}{x} \, dx$ (i) $\int_{0}^{\pi} -\sin x \, dx$ (j) $\int_{1}^{4} 13 \, dx$

(k) $\int_{0}^{3} x^2 + \sin x \, dx$ (l) $\int_{1}^{3} 5x^2 \, dx$

(m) $\int_{0}^{1} (-x) \, dx$ (n) $\int_{1}^{3} x^{-2} \, dx$

(o) $\int_{0}^{1} \sin 3t \, dt$

22.5   ## Areas under curves

As we stated at the beginning of this chapter integration can be used to find the area underneath a curve and above the x axis. Consider the shaded area in Figure 22.3 which lies under the curve $y = f(x)$, above the x axis, between the values $x = a$ and $x = b$. Provided all of the required area lies above the x axis, this area is given by the following formula.

KEY POINT

$$\text{area} = \int_{a}^{b} f(x) \, dx$$

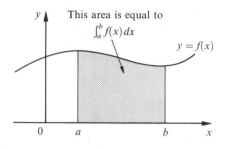

Figure 22.3
The shaded area is given
by $\int_a^b f(x)dx$

Worked examples

22.13   Find the area under $y = 7x + 3$ between $x = 0$ and $x = 2$.

Solution   This area is shown in Figure 22.4. To find this area we must evaluate
the definite integral $\int_0^2 7x + 3\,dx$:

$$\text{area} = \int_0^2 7x + 3\,dx = \left[\frac{7x^2}{2} + 3x\right]_0^2$$

$$= \left\{\frac{7(2)^2}{2} + 3(2)\right\} - (0)$$

$$= 20$$

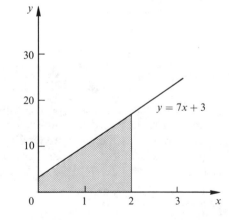

Figure 22.4
The shaded area is given
by $\int_0^2 7x + 3\,dx$

The required area is 20 square units. Convince yourself that this is
correct by calculating the area of the shaded trapezium using the
formula given in Chapter 20.

22.14   Find the area under the curve $y = x^2$ and above the x axis, between $x = 2$ and $x = 5$.

Solution   The required area is shown in Figure 22.5.

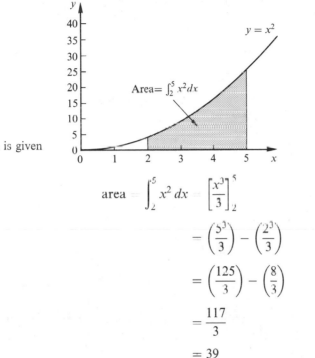

Figure 22.5
The shaded area is given by $\int_2^5 x^2 \, dx$

$$\text{area} = \int_2^5 x^2 \, dx = \left[\frac{x^3}{3} \right]_2^5$$

$$= \left(\frac{5^3}{3} \right) - \left(\frac{2^3}{3} \right)$$

$$= \left(\frac{125}{3} \right) - \left(\frac{8}{3} \right)$$

$$= \frac{117}{3}$$

$$= 39$$

The required area is 39 square units.

Note that the definite integral $\int_a^b f(x)dx$ only represents the area under $y = f(x)$ between $x = a$ and $x = b$ when all this area lies above the x axis. If between a and b the curve dips below the x axis, the result obtained will not be the area bounded by the curve. We shall see how to overcome this now.

Finding areas when some or all lie below the x axis

Consider the graph of $y = x^3$ shown in Figure 22.6. Suppose we want to know the area enclosed by the graph and the x axis between $x = -1$ and $x = 1$. Evaluating $\int_{-1}^1 x^3 \, dx$ we find:

$$\int_{-1}^{1} x^3 \, dx = \left[\frac{x^4}{4} \right]_{-1}^{1}$$

$$= \left\{ \frac{(1)^4}{4} \right\} - \left\{ \frac{(-1)^4}{4} \right\}$$

$$= \frac{1}{4} - \frac{1}{4}$$

$$= 0$$

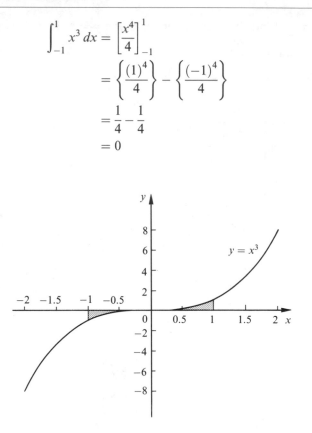

Figure 22.6
Graph of $y = x^3$

The result of finding the definite integral is 0, even though there is clearly a region enclosed by the curve and the x axis. The reason for this is that part of the curve lies below the x axis. Let us evaluate two separate integrals, one representing the area below the x axis and the other representing the area above the x axis:

$$\int_{-1}^{0} x^3 \, dx = \left[\frac{x^4}{4} \right]_{-1}^{0}$$

$$= \left\{ \frac{(0)^4}{4} \right\} - \left\{ \frac{(-1)^4}{4} \right\}$$

$$= 0 - \left(\frac{1}{4} \right)$$

$$= -\frac{1}{4}$$

The result is negative, that is, $-\frac{1}{4}$. This means that the area bounded by the curve and the x axis between $x = -1$ and $x = 0$ is equal to $\frac{1}{4}$, the negative sign indicating that this area is below the x axis. Also:

$$\int_0^1 x^3\, dx = \left[\frac{x^4}{4}\right]_0^1$$

$$= \left\{\frac{(1)^4}{4}\right\} - \left\{\frac{(0)^4}{4}\right\}$$

$$= \frac{1}{4}$$

So the area between $x = 0$ and $x = 1$ is $\frac{1}{4}$. Note that we could have deduced this from the symmetry in the graph. The total area between $x = -1$ and $x = 1$ is therefore equal to $\frac{1}{4} + \frac{1}{4} = \frac{1}{2}$. The important point to note is that when some or all of the area lies below the x axis it is necessary to find the total area by finding the area of each part. It is therefore essential to draw a graph of the function to be integrated first.

Self-assessment questions 22.5

1. Explain how the area bounded by a curve and the x axis can be found using a definite integral. What precautions must be taken when doing this?

Exercise 22.5

1. Find the area bounded by the curve $y = x^3 + 3$, the x axis and the lines $x = 1$ and $x = 3$.

2. Find the area bounded by the curve $y = 4 - x^2$ and the x axis between $x = -1$ and $x = 4$.

3. Find the area under $y = e^{2t}$ from $t = 1$ to $t = 4$.

4. Find the area under $y = 1/t$ for $1 \le t \le 5$.

5. Find the area under $y = \sqrt{x}$ between $x = 1$ and $x = 2$.

Test and assignment exercises 22

1. Find the following indefinite integrals:

(a) $\int 3x\, dx$ (b) $\int 2x + 4\, dx$ (c) $\int x^2 + 7\, dx$

(d) $\int 3x^2 + 7x - 7\, dx$ (e) $\int e^{1.5x}\, dx$

(f) $\int \sin 5x - \cos 4x\, dx$ (g) $\int 5e^{-x}\, dx$

2. Evaluate the following definite integrals:

(a) $\int_0^4 x^2\, dx$ (b) $\int_0^4 x^4\, dx$ (c) $\int_{-1}^2 3x^3\, dx$

(d) $\int_0^1 x^2 + 3x + 4\, dx$ (e) $\int_{-2}^2 x^3 + 2\, dx$

(f) $\int_{-3}^2 2x - 13\, dx$ (g) $\int_3^5 x + 2x + 3x^2\, dx$

(h) $\int_2^7 5\, dx$ (i) $\int_1^2 11x^4\, dx$

3. Calculate the area bounded by the curve $y = x^2 - 4x + 3$ and the x axis from $x = 1$ to $x = 3$.

4. Sketch a graph of $y = \sin x$ from $x = 0$ to $x = \pi$. Calculate the area bounded by your curve and the x axis.

5. Find $\int 4e^{2t}\, dt$.

6. Find $\int 5 + x + x^2 + \dfrac{5}{x} dx$

7. Find $\int_1^2 \dfrac{1}{x^2} dx$

8. Calculate the area bounded by $y = e^{-x}$ and the x axis between

(a) $x = 0$ and $x = 1$ (b) $x = 1$ and $x = 2$
(c) $x = 0$ and $x = 2$

Deduce that

$$\int_0^2 e^{-x} dx = \int_0^1 e^{-x} dx + \int_1^2 e^{-x} dx$$

23 Tables and charts

Objectives

This chapter

- explains the distinction between discrete and continuous data
- shows how raw data can be organized using a tally chart
- explains what is meant by a frequency distribution and a relative frequency distribution
- shows how data can be represented in the form of bar charts, pie charts, pictograms and histograms

23.1 Introduction to data

In the modern world, information from a wide range of sources is gathered, presented and interpreted. Most newspapers, television news programmes and documentaries contain vast numbers of facts and figures concerning almost every aspect of life, including environmental issues, the lengths of hospital waiting lists, crime statistics and economic performance indicators. Information such as this, which is gathered by carrying out surveys and doing research, is known as **data**.

Sometimes data must take on a value from a specific set of numbers, and no other values are possible. For example, the number of children in a family must be 0, 1, 2, 3, 4 and so on, and no intermediate values are allowed. It is impossible to have 3.32 children, say. Usually one allowable value differs from the next by a fixed amount. Such data is said to be **discrete**. Other examples of discrete data include

- the number of car thefts in a city in one week – this must be 0, 1, 2, 3, and so on. You cannot have three and a half thefts.

- shoe sizes – these can be $\dots, 3\frac{1}{2}, 4, 4\frac{1}{2}, 5, 5\frac{1}{2}, 6, 6\frac{1}{2}, 7$ and so on. You cannot have a shoe size of 9.82.

Sometimes data can take on *any* value within a specified range. Such data is called **continuous**. The volume of liquid in a 1 litre jug can take any value from 0 to 1 litre. The lifespan of an electric light bulb could be any non-negative value.

Frequently data is presented in the form of tables and charts, the intention being to make the information readily understandable. When data is first collected, and before it is processed in any way, it is known as **raw data**. For example, in a test the mathematics marks out of 10 of a group of 30 students are

$$7 \quad 5 \quad 5 \quad 8 \quad 9 \quad 10 \quad 9 \quad 10 \quad 7 \quad 8 \quad 6 \quad 3 \quad 5 \quad 9 \quad 6$$

$$10 \quad 8 \quad 8 \quad 7 \quad 8 \quad 6 \quad 7 \quad 8 \quad 8 \quad 10 \quad 9 \quad 4 \quad 5 \quad 9 \quad 8$$

This is raw data. To try to make sense of this data it is helpful to find out how many students scored each particular mark. This can be done by means of a **tally chart.** To produce this we write down in a column all the possible marks, in order. We then go through the raw data and indicate the occurrence of each mark with a vertical line or tally like |. Every time a fifth tally is recorded this is shown by striking through the previous four, as in ⊬ϯϯϯ. This makes counting up the tallies particularly easy. The tallies for all the marks are shown in Table 23.1. The number of occurrences of each mark is called its **frequency** and the frequencies can be found from the number of tallies.We see that the tally chart is a useful way of organizing the raw data into a form which will enable us to answer questions and obtain useful information about it.

Table 23.1
Tally chart for mathematics marks

Mark	Tally	Frequency
0		0
1		0
2		0
3	I	1
4	I	1
5	IIII	4
6	III	3
7	IIII	4
8	⊬ϯϯϯ III	8
9	⊬ϯϯϯ	5
10	IIII	4
Total		30

Nowadays much of the tedium of producing tally charts for very large amounts of data can be avoided using computer programs available specifically for this purpose.

Self-assessment questions 23.1

1. Explain the distinction between discrete data and continuous data. Give a new example of each.
2. Explain the purpose of a tally chart.

Exercise 23.1

1. State whether each of the following is an example of discrete data or continuous data:

   (a) the number of matches found in a matchbox
   (b) the percentage of sulphur dioxide found in an air sample taken above a city
   (c) the number of peas in a packet of frozen peas
   (d) the radius of a ball bearing
   (e) the weight of a baby
   (f) the number of students in a particular type of accommodation
   (g) the weight of a soil sample
   (h) the temperature of a patient in hospital
   (i) the number of telephone calls received by an answering machine during one day

2. Twenty-five students were asked to state their year of birth. The information given was

1975	1975	1974	1974	1975	1976
1976	1974	1972	1973	1973	1970
1975	1975	1973	1970	1975	1974
1975	1974	1970	1973	1974	1974
1975					

   Produce a tally chart to organize this data and state the frequency of occurrence of each year of birth.

23.2   Frequency tables and distributions

Data is often presented in a **table**. For example, in a recent survey, 1000 students were asked to state the type of accommodation in which they lived during term time. The data gathered is given in Table 23.2. As we saw from the work in the previous section, the number of occurrences of each entry in the table is known as its **frequency**. So, for example, the frequency of students in private rented accommodation is 250. The table summarizes the various frequencies and is known as a **frequency distribution.**

A **relative frequency distribution** is found by expressing each frequency as a proportion of the total frequency.

Table 23.2
Frequency distribution

Type of accommodation	Number of students
Hall of residence	675
Parental home	50
Private rented	250
Other	25
Total	1000

KEY POINT

The relative frequency is found by dividing a frequency by the total frequency.

In this example, because the total frequency is 1000, the relative frequency is easy to calculate. The number of students living at their parental home is 50 out of a total of 1000. Therefore the relative frequency of this group of students is $\frac{50}{1000}$ or 0.050. Table 23.3 shows the relative frequency distribution of student accommodation. Note that the relative frequencies always sum to 1.

Table 23.3
Relative frequency distribution

Type of accommodation	Number of students	Relative frequency
Hall of residence	675	0.675
Parental home	50	0.050
Private rented	250	0.250
Other	25	0.025
Total	1000	1.000

It is sometimes helpful to express the frequencies as percentages. This is done by multiplying each relative frequency by 100, which gives the results shown in Table 23.4. We see from this that 5% of students questioned live at their parental home.

KEY POINT

A frequency can be expressed as a percentage by multiplying its relative frequency by 100.

Table 23.4 Frequencies expressed as percentages	Type of accommodation	Number of students	Relative frequency × 100
	Hall of residence	675	67.5%
	Parental home	50	5.0%
	Private rented	250	25.0%
	Other	25	2.5%
	Total	1000	100%

Worked example

23.1 In 1988 there were 4320 areas throughout the world which were designated as national parks. Table 23.5 is a frequency distribution showing how these parks are distributed in different parts of the world.

Table 23.5 The number of national parks throughout the world	Region	Number of national parks
	Africa	486
	N. America	587
	S. America	315
	Former Soviet Union	168
	Asia	960
	Europe	1032
	Oceania	767
	Antarctica	5
	Total	4320

(a) Express this data as a relative frequency distribution and also in terms of percentages.

(b) What percentage of the world's national parks are in Africa?

Solution (a) The relative frequencies are found by expressing each frequency as a proportion of the total. For example, Europe has 1032 national parks out of a total of 4320, and so its relative frequency is $\frac{1032}{4320} = 0.2389$. Figures for the other regions have been calculated and are shown in Table 23.6. To express a relative frequency as a

Table 23.6
The number of national parks throughout the world

Region	Number	Relative frequency	Rel. freq. ×100
Africa	486	0.1125	11.25%
N. America	587	0.1359	13.59%
S. America	315	0.0729	7.29%
Former Soviet Union	168	0.0389	3.89%
Asia	960	0.2222	22.22%
Europe	1032	0.2389	23.89%
Oceania	767	0.1775	17.75%
Antarctica	5	0.0012	0.12%
Total	4320	1.000	100%

percentage it is multiplied by 100. Thus we find that Europe has 23.89% of the world's national parks. The percentages for the other regions are also shown in Table 23.6.

(b) We note from Table 23.6 that 11.25% of the world's national parks are in Africa.

When dealing with large amounts of data it is usual to group the data into classes. For example, in an environmental experiment the weights in grams of 30 soil samples were measured using a balance. We could list all 30 weights here but such a long list of data is cumbersome. Instead, we have already grouped the data into several weight ranges or **classes**. Table 23.7 shows the number of samples in each class. Such a table is called a **grouped frequency distribution.**

Table 23.7
Grouped frequency distribution

Weight range (grams)	Number of samples in that range
100–109	5
110–119	8
120–129	11
130–139	5
140–149	1
Total	30

A disadvantage of grouping the data in this way is that information about individual weights is lost. However, this disadvantage is usually outweighed by having a more compact set of data which is easier to work with. There are five samples which lie in the first class, eight samples in the second and so on. For the first class, the numbers 100 and 109 are called **class limits**, the smaller number being the **lower class limit** and the larger being the **upper class limit**. Theoretically, the class 100 109 includes all weights from 99.5 g up to but not including 109.5 g. These numbers are called the **class boundaries**. In practice the class boundaries are found by adding the upper class limit of one class to the lower limit of the next class and dividing by 2. The **class width** is the difference between the larger class boundary and the smaller. The class 100–109 has width $109.5 - 99.5 = 10$. We do not know the actual values of the weights in each class. Should we require an estimate, the best we can do is use the value in the middle of the class, known as the class **midpoint**. The midpoint of the first class is 104.5; the midpoint of the second class is 114.5 and so on. The midpoint can be found by adding half the class width to the lower class boundary.

Worked example

23.2   The blood pressure in millimetres of mercury of 20 workers was recorded to the nearest millimetre. The data collected was as follows:

>121   123   124   129   130   119   129   124   119   121   122
>
>124   124   128   129   136   120   119   121   136

(a) Group this data into classes 115–119, 120–124, and so on.

(b) State the class limits of the first three classes.

(c) State the class boundaries of the first three classes.

(d) What is the class width of the third class?

(e) What is the midpoint of the third class?

Solution   (a) To group the data a tally chart is used as shown in Table 23.8.

(b) The class limits of the first class are 115 and 119; those of the second class are 120 and 124; those of the third are 125 and 129.

(c) Theoretically, the class 115–119 will contain all blood pressures between 114.5 and 119.5 and so the class boundaries of the first class are 114.5 and 119.5. The class boundaries of the second class are 119.5 and 124.5, and those of the third class are 124.5 and 129.5.

Table 23.8
Tally chart for blood
pressures

Pressure (mm mercury)	Tally	Frequency
115–119	|||	3
120–124	ⅢⅢ ⅢⅢ	10
125–129	||||	4
130–134	|	1
135–139	||	2
Total		20

(d) The upper class boundary of the third class is 129.5. The lower class boundary is 124.5. The difference between these values gives the class width, that is, $129.5 - 124.5 = 5$.

(e) The midpoint of the third class is found by adding half the class width to the lower class boundary, that is, $2.5 + 124.5 = 127$.

Self-assessment questions 23.2

1. Why is it often useful to present data in the form of a *grouped* frequency distribution?
2. Explain the distinction between class boundaries and class limits.
3. How is the class width calculated?
4. How is the class midpoint calculated?

Exercise 23.2

1. The age of each patient over the age of 14 visiting a doctors' practice in one day was recorded as follows:

18	18	76	15	72	45	48	62
21	27	45	43	28	19	17	37
35	34	23	25	46	56	32	18
23	34	32	56	29	43		

(a) Use a tally chart to produce a grouped frequency distribution with age groupings 15–19, 20–24, 25–29, and so on.
(b) What is the relative frequency of patients in the age group 70–74?
(c) What is the relative frequency of patients aged 70 and over?
(d) Express the number of patients in the age group 25–29 as a percentage.

2. Consider the following table which shows the lifetimes, to the nearest hour, of 100 energy saving light bulbs.

(a) If a light bulb has a life of 5499.4 hours into which class will it be put?
(b) If a light bulb has a life of 5499.8 hours into which class will it be put?
(c) State the class boundaries of each class.

(d) Find the class width.
(e) State the class midpoint of each class.

Lifetime (hours)	Frequency
5000–5499	6
5500–5999	32
6000–6499	58
6500–6999	4
Total	100

3. The percentages of 60 students in an Information Technology test are given as follows.

```
45  92  81  76  51  46  82  65  61
19  62  58  72  65  66  97  61  57
63  93  61  46  47  61  56  45  39
47  55  58  81  71  52  38  53  59
82  92  87  86  51  29  19  79  55
18  53  29  87  87  77  85  67  89
17  29  86  57  59  57
```

Form a frequency distribution by drawing up a tally chart using class intervals 0–9, 10–19 and so on.

4. The data in the table below gives the radius of 20 ball bearings. State the class boundaries of the second class and find the class width.

Radius (mm)	Frequency
20.56–20.58	3
20.59–20.61	6
20.62 20.64	8
20.65–20.67	3

23.3   Bar charts, pie charts, pictograms and histograms

In a **bar chart** information is represented by rectangles or bars. The bars may be drawn horizontally or vertically. The length of each bar corresponds to a frequency.

The data given earlier in Table 23.2 concerning student accommodation is presented in the form of a horizontal bar chart in Figure 23.1. Note that the scale on the horizontal axis must be uniform, that is, it must be evenly spaced, and that the type of accommodation is clearly identified on each bar. The bar chart must be given a title to explain the information that is being presented.

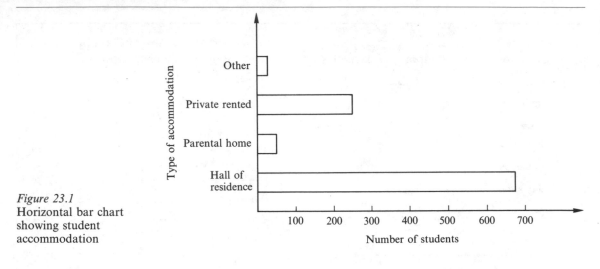

Figure 23.1
Horizontal bar chart
showing student
accommodation

Worked example

23.3 A reproduction antique furniture manufacturer has been producing pine wardrobes for the past six years. The number of wardrobes sold each year is given in Table 23.9. Produce a vertical bar chart to illustrate this information.

Solution The information is presented in the form of a vertical bar chart in Figure 23.2.

Table 23.9
Wardrobes sold

Year	Number sold
1987	15
1988	16
1989	20
1990	20
1991	24
1992	30

Figure 23.2
Vertical bar chart
showing the number of
wardrobes sold each year

In a **pie chart**, a circular 'pie' is divided into a number of portions, with each portion, or **sector**, of the pie representing a different category. The whole pie represents all categories together. The size of a particular portion must represent the number in its category and this is done by dividing the circle proportionately. Usually a protractor will be required.

Worked example

23.4   A sample of 360 people were asked to state their favourite flavour of potato crisp: 75 preferred 'plain' crisps, 120 preferred 'cheese and onion', 90 preferred 'salt and vinegar' and the remaining 75 preferred 'beef'. Draw a pie chart to present this data.

Solution   All categories together, that is, all 360 people asked, make up the whole pie. Because there are 360° in a circle this makes the pie chart particularly easy to draw. A sector of angle 75° will represent 'plain crisps', a sector of 120° will represent 'cheese and onion' and so on. The pie chart is shown in Figure 23.3.

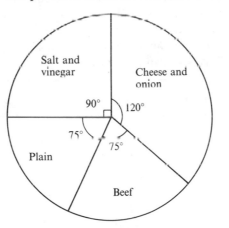

Figure 23.3
Pie chart showing preferred flavour of crisps of 360 people

The pie chart in the previous example was simple to draw because the total number of all categories was 360, the same as the number of degrees in a circle. Let us now see how to deal with a situation where this is not the case. First each relative frequency is found and then this is multiplied by 360. The result is the angle, in degrees, of the corresponding sector.

KEY POINT

The angle, in degrees, of each sector in a pie chart is found by multiplying the corresponding relative frequency by 360. This gives the angle of each sector.

Worked example

23.5   Five hundred students were asked to state how they usually travelled to and from their place of study. The results are given in Table 23.10. Present this information in a pie chart.

Table 23.10
Means of travel

Means of travel	Frequency f
Bus	50
Walk	180
Cycle	200
Car	40
Other	30
Total	500

Solution   To find the angle of each sector of the pie chart we first find the corresponding relative frequencies. Recall that this is done by dividing each frequency by the total, 500. Then each relative frequency is multiplied by 360. The results are shown in Table 23.11. The table gives the required angles. Using a protractor the pie chart can then be drawn. This is shown in Figure 23.4.

Table 23.11
Means of travel

Means of travel	Frequency f	Rel. freq. $f/500$	Rel. freq. $\times 360$
Bus	50	$\frac{50}{500}$	$\frac{50}{500} \times 360 = 36.0°$
Walk	180	$\frac{180}{500}$	$\frac{180}{500} \times 360 = 129.6°$
Cycle	200	$\frac{200}{500}$	$\frac{200}{500} \times 360 = 144.0°$
Car	40	$\frac{40}{500}$	$\frac{40}{500} \times 360 = 28.8°$
Other	30	$\frac{30}{500}$	$\frac{30}{500} \times 360 = 21.6°$
Total	500		$360°$

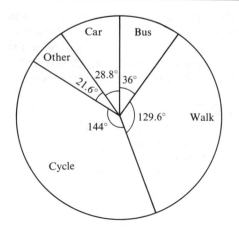

Figure 23.4
Pie chart showing how students travel to work

A **pictogram** uses a picture to represent data in an eye-catching way. There are many ways to produce pictograms but generally the number of objects drawn represents the number of items in a particular category. This is a common form of representation popular on television when the viewer only sees the information for a few seconds.

Worked example

23.6   The number of hours of sunshine in the month of August 1975 in four popular holiday resorts is given in Table 23.12. Show this information using a pictogram. Use one 'sun' to represent 100 hours of sunshine.

Table 23.12
Hours of sunshine in August 1975 in four resorts

Resort	Number of hours
Torbay	300
Scarborough	275
Blackpool	350
Brighton	325

Solution   The pictogram is shown in Figure 23.5.

Figure 23.5
Pictogram showing hours of sunshine in August 1975

A **histogram** is often used to illustrate data which is continuous or has been grouped. It is similar to a vertical bar chart in that it is drawn by constructing vertical rectangles. However, in a histogram it is the *area* of each rectangle, and not its length, which is proportional to the frequency of each class. When all class widths are the same this point is not important and you can assume that the length of a rectangle represents the frequency. Class boundaries and not class limits must be used on the horizontal axis to distinguish one class from the next. Consider the following example.

Worked example

23.7 The heights, to the nearest centimetre, of 100 students are given in Table 23.13.

(a) State the class limits of the second class.

(b) Identify the class boundaries of the first class and the last class.

(c) In which class would a student whose actual height is 167.4 cm be placed?

(d) In which class would a student whose actual height is 167.5 cm be placed?

(e) Draw a histogram to depict this data.

Table 23.13
Height of 100 students

Height (cm)	Frequency
164–165	4
166–167	8
168–169	10
170–171	27
172–173	30
174–175	10
176–177	6
178–179	5
Total	100

Solution (a) The lower class limit of the second class is 166 cm. The upper class limit is 167 cm.

(b) The class boundaries of the first class are 163.5 cm and 165.5 cm. The class boundaries of the last class are 177.5 cm and 179.5 cm.

(c) A student with height 167.4 cm will have his height recorded to the nearest centimetre as 167. He will be placed in the class 166–167.

(d) A student with height 167.5 cm will have her height recorded to the nearest centimetre as 168. She will be placed in the class 168–169.

(e) The histogram is shown in Figure 23.6. Note that the class boundaries are used on the horizontal axis to distinguish one class from the next. The area of each rectangle is proportional to the frequency of each class.

Figure 23.6
Histogram showing the
distribution of heights of
100 students

Self-assessment questions 23.3

1. There is an important distinction between a histogram and a vertical bar chart. Can you explain this?
2. Give one advantage and one disadvantage of using a pictogram to represent data.
3. On a histogram the class boundaries are used on the horizontal axis to distinguish one class from the next. Why do you think class boundaries as opposed to class limits are used?

Exercise 23.3

1. A local council wants to produce a leaflet explaining how the council tax is spent. For every one pound spent, 40 p went to Education, 15 p went to the Police, 15 p went to Cleansing and Refuse, 20 p went to Highways and the remaining 10 p went to a variety of other services. Show this information

   (a) using a pie chart, and (b) using a pictogram.

2. The composition of the atmosphere of the Earth is 78% nitrogen, 21% oxygen and 1% other gases. Draw a pie chart to show this information.

3. An insurance company keeps records on the distribution of the amount of small claims for theft it receives. Amounts are given to the nearest pound. This information is given in the following table.

Amount (£)	Frequency
0–99	5
100–199	8
200–299	16
300–399	24
400–499	26

(a) State the class boundaries.

(b) In which class would a claim for £199.75 be placed?

(c) Draw a histogram to show this information.

Test and assignment exercises 23

1. State whether the following are examples of discrete or continuous data:

   (a) the temperature of a furnace

   (b) the number of goals scored in a football match

   (c) the amount of money in a worker's wage packet at the end of the week.

2. A group of students were interviewed to obtain information on their termly expenditure. On average, for every pound spent, 60p was spent on accommodation, fuel, etc., 25p was spent on food, 10p was spent on entertainment and the remainder on a variety of other items. Draw a pie chart to show this expenditure.

3. Forty lecturers were asked to state which newspaper they bought most regularly. The results were: *Guardian*, 15; *Independent*, 8; *Times*, 2; *Daily Express*, 7; *Mail*, 2; None, 6. Produce a pie chart to show this data.

4. The percentage of total body weight formed by various parts of the body of a newborn baby is as follows: muscles, 26%; skeleton, 17%; skin, 20%; heart, lungs etc., 11%; remainder, 26%. Depict this information (a) on a pie chart, (b) using a bar chart.

5. The following marks were obtained by 27 students in a test. Draw up a frequency distribution by means of a tally chart.

5	4	6	2	9	1	10	10	5	5	6
5	5	9	8	7	9	6	7	6	8	7
6	9	8	8	7						

6. The number of cars sold by a major manufacturer of luxury cars in the last 10 years is given in the table below. Show this data on a horizontal bar chart.

Year	Number
1983	2000
1984	2320
1985	2380
1986	2600
1987	2650
1988	2750
1989	2750
1990	2800
1991	1500
1992	1425

7. The table below shows the frequency distribution of the lifetimes in hours of 100 light bulbs tested.

   (a) State the class limits of the first and second classes.

   (b) State the class boundaries of all classes.

Lifetime (hours)	Frequency
600–699	13
700–799	25
800–899	32
900–999	10
1000–1099	15
1100–1199	5

(c) Determine each class width and midpoint.

(d) Draw a histogram for this frequency distribution.

8. In each year from 1988 to 1993 the numbers of visitors to a museum were respectively 12000, 13000, 14500, 18000, 18000 and 17500. Show this information in a pictogram.

24 Statistics

Objectives

This chapter

- introduces the three common averages – mean, median and mode – and shows how to calculate them

- explains what is meant by the variance and standard deviation and shows how to calculate them

24.1 Introduction

In the previous chapter we outlined how data gathered from a range of sources can be represented. In order to get the most out of such data and to be able to interpret it sensibly and reliably, techniques have been developed for its analysis. **Statistics** is the name given to this science. One important statistical quantity is an 'average', the purpose of which is to represent a whole set of data by a single number which in some way represents the set. This chapter explains three types of average and shows how to calculate them. Finally the 'variance' and 'standard deviation' are introduced. These are numbers which describe how widely the data is spread.

24.2 Averages: the mean, median and mode

When presented with a large amount of data it is often useful to ask 'is there a single number which typifies the data?' For example, following an examination taken by a large number of candidates, examiners may be dealing with hundreds of examination scripts. The marks on these will vary from those with very poor marks to those with very high or even top marks. In order to judge whether the group of students as a whole found the examination difficult, the examiner may be asked to provide a single value which gives a measure of how well the students have performed. Such a value is called an **average.** In statistics there are three important averages: the arithmetic mean, the median and the mode.

The arithmetic mean

The **arithmetic mean,** or simply the **mean,** of a set of values is found by adding up all the values and dividing the result by the total number of values in the set. It is given by the following formula.

KEY POINT

$$\text{mean} = \frac{\text{sum of the values}}{\text{total number of values}}$$

Worked examples

24.1 Eight students sit a mathematics test and their marks out of ten are 4, 6, 6, 7, 7, 7, 8, and 10. Find the mean mark.

Solution The sum of the marks is $4 + 6 + 6 + 7 + 7 + 7 + 8 + 10 = 55$. The total number of values is 8. Therefore,

$$\text{mean} = \frac{\text{sum of the values}}{\text{total number of values}} = \frac{55}{8} = 6.875$$

The examiner can quote 6.875 out of 10 as the 'average mark' of the group of students.

24.2 In a hospital a patient's body temperature is recorded every hour for six hours. Find the mean temperature over the six-hour period if the six temperatures, in °C, were

36.5   36.8   36.9   36.9   36.9   37.0

Solution To find the mean temperature the sum of all six values is found and the result is divided by 6. That is,

$$\text{mean} = \frac{36.5 + 36.8 + 36.9 + 36.9 + 36.9 + 37.0}{6} = \frac{221.0}{6} = 36.83\,°\text{C}$$

In more advanced work we make use of a formula for the mean which requires knowledge of some special notation. Suppose we have n values and we call these x_1, x_2, x_3 and so on up to x_n. The mean of these values is given the symbol $\bar{x}$, pronounced 'x bar'. To calculate the mean we must add up these values and divide by n, that is,

$$\text{mean} = \bar{x} = \frac{x_1 + x_2 + x_3 + \cdots + x_n}{n}$$

A notation is often used to shorten this formula. In mathematics, the Greek letter sigma, written $\sum$, stands for a sum. The sum

$$x_1 + x_2 \text{ is written } \sum_{i=1}^{2} x_i$$

and the sum

$$x_1 + x_2 + x_3 + x_4 \text{ is written } \sum_{i=1}^{4} x_i$$

Note that i runs through all integer values from 1 to n.

where the values above and below the sigma sign give the first and last values in the sum. Similarly $x_1 + x_2 + x_3 + \cdots + x_n$ is written $\sum_{i=1}^{n} x_i$. Using this notation, the formula for the mean can be written in the following way:

KEY POINT

$$\text{mean} = \bar{x} = \frac{\sum_{i=1}^{n} x_i}{n}$$

Worked examples

24.3   Express the following in sigma notation:

(a) $x_1 + x_2 + x_3 + \cdots + x_8 + x_9$ (b) $x_{10} + x_{11} + \cdots + x_{100}$

Solution   (a) $x_1 + x_2 + x_3 + \cdots + x_8 + x_9 = \sum_{i=1}^{9} x_i$

(b) $x_{10} + x_{11} + \cdots + x_{100} = \sum_{i=10}^{100} x_i$

24.4   Find the mean of the values $x_1 = 5$, $x_2 = 7$, $x_3 = 13$, $x_4 = 21$ and $x_5 = 29$.

Solution   The number of values equals 5, so we let $n = 5$. The sum of the values is

$$\sum_{i=1}^{5} x_i = x_1 + x_2 + x_3 + x_4 + x_5 = 5 + 7 + 13 + 21 + 29 = 75$$

The mean is

$$\text{mean} = \bar{x} = \frac{\sum_{i=1}^{5} x_i}{n} = \frac{75}{5} = 15$$

When the data is presented in the form of a frequency distribution the mean is found by first multiplying each data value by its frequency. The results are added. This is equivalent to adding up all the data values. The mean is found by dividing this sum by the sum of all the frequencies. Note that the sum of the frequencies is equal to the total number of values. Consider the following example.

Worked example

24.5   Thirty-eight students sit a mathematics test and their marks out of 10 are shown in Table 24.1. Find the mean mark.

Mark, m	Frequency, f
0	0
1	0
2	1
3	0
4	1
5	7
6	16
7	8
8	3
9	1
10	1
Total	38

Table 24.1
Marks of 38 students in a test

Mark, m	Frequency, f	$m \times f$
0	0	0
1	0	0
2	1	2
3	0	0
4	1	4
5	7	35
6	16	96
7	8	56
8	3	24
9	1	9
10	1	10
Totals	38	236

Table 24.2
Marks of 38 students multiplied by frequency

Solution   Each data value, in this case the mark, is multiplied by its frequency, and the results are added. This is equivalent to adding up all the 38 individual marks. This is shown in Table 24.2.

Note that the sum of all the frequencies is equal to the number of students taking the test. The number 236 is equal to the sum of all the individual marks. The mean is found by dividing this sum by the sum of all the frequencies:

$$\text{mean} = \frac{236}{38} = 6.21$$

The mean mark is 6.21 out of 10.

Using the sigma notation the formula for the mean mark of a frequency distribution with N classes, where the frequency of value x_i is f_i, becomes:

KEY POINT

$$\text{mean} = \bar{x} = \frac{\sum_{i=1}^{N} f_i \times x_i}{\sum_{i=1}^{N} f_i}$$

Note that $\sum_{i=1}^{N} f_i = n$, that is, the sum of all the frequencies equals the total number of values.

When the data is in the form of a grouped distribution the class midpoint is used to calculate the mean. Consider the following example.

Worked example

24.6   The heights, to the nearest centimetre, of 100 students are given in Table 24.3. Find the mean height.

Table 24.3
Heights of 100 students

Height (cm)	Frequency
164–165	4
166–167	8
168–169	10
170–171	27
172–173	30
174–175	10
176–177	6
178–179	5
Total	100

Solution   Because the actual heights of students in each class are not known we use the midpoint of the class as an estimate. The midpoint of the class 164–165 is 164.5. Other midpoints and the calculation of the mean are shown in Table 24.4.

Then

$$\text{mean} = \bar{x} = \frac{\sum_{i=1}^{N} f_i \times x_i}{\sum_{i=1}^{N} f_i} = \frac{17150}{100} = 171.5$$

The mean height is 171.5 cm.

Table 24.4
Heights of 100 students
with midpoints multiplied
by frequency

Height (cm)	Frequency f_i	Midpoint x_i	$f_i \times x_i$
164–165	4	164.5	658.0
166–167	8	166.5	1332.0
168–169	10	168.5	1685.0
170–171	27	170.5	4603.5
172–173	30	172.5	5175.0
174–175	10	174.5	1745.0
176–177	6	176.5	1059.0
178–179	5	178.5	892.5
Total	100		17150

The median

A second average which also typifies a set of data is the **median**.

KEY POINT

> The **median** of a set of numbers is found by listing the numbers in ascending order and then selecting the value that lies half-way along the list.

Worked example

24.7   Find the median of the numbers

   1   2   6   7   9   11   11   11   14

Solution   The set of numbers is already given in order. The number half way along the list is 9, because there are four numbers before it and four numbers after it in the list. Hence the median is 9.

When there is an even number of values, the median is found by taking the mean of the two middle values.

Worked example

24.8   Find the median of the following salaries: £24000, £12000, £16000, £22000, £10000 and £25000.

Solution   The numbers are first arranged in order as £10000, £12000, £16000, £22000, £24000 and £25000. Because there is an even number of values there are two middle figures: £16000 and £22000. The mean of these is

$$\frac{16000 + 22000}{2} = 19000$$

The median salary is therefore £19000.

The mode

A third average is the **mode**.

KEY POINT

> The **mode** of a set of values is that value which occurs most often.

Worked examples

24.9   Find the mode of the set of numbers

1   1   4   4   5   6   8   8   8   9

Solution   The number which occurs most often is 8, which occurs three times. Therefore 8 is the mode. Usually a mode is quoted when we want to represent the most popular value in a set.

Sometimes a set of data may have more than one mode.

24.10   Find the mode of the set of numbers

20   20   21   21   21   48   48   49   49   49

Solution   In this example there is no single value which occurs most frequently. The number 21 occurs three times, but so does the number 49. This set has two modes. The data is said to be **bimodal**.

Self-assessment questions 24.2

1. State the three different types of averages commonly used in statistical calculations.
2. In an annual report, an employer of a small firm claims that the median salary for his workforce is £18500. However, over discussions in the canteen it is apparent that no worker earns this amount. Explain how this might have arisen.

Exercise 24.2

1. Find the mean, median and mode of the following set of values: 2, 3, 5, 5, 5, 5, 8, 8, 9.

2. Find the mean of the set of numbers 1, 1, 1, 1, 1, 1, 256. Explain why the mean does not represent the data adequately. Which average might it have been more appropriate to use?

3. The marks of seven students in a test were 45%, 83%, 99%, 65%, 68%, 72% and 66%. Find the mean mark and the median mark.

4. Write out fully each of the following expressions:

   (a) $\sum_{i=1}^{4} x_i$ (b) $\sum_{i=1}^{7} x_i$

   (c) $\sum_{i=1}^{3}(x_i - 3)^2$ (d) $\sum_{i=1}^{4}(2 - x_i)^2$

5. Write the following concisely using sigma notation:

   (a) $x_3 + x_4 + x_5 + x_6$
   (b) $(x_1 - 1) + (x_2 - 1) + (x_3 - 1)$

6. Calculate the mean, median and mode of the following numbers: 1.00, 1.15, 1.25, 1.38, 1.39 and 1.40.

7. The prices of the eight terrace houses advertised by a local estate agent are

   £29000   £37500   £32500   £29995

   £31995   £32750   £29950   £32950

   Find the mean price of these houses.

8. The data in the table gives the radius of 20 ball bearings. Find the class midpoints and hence calculate the average radius.

Radius (mm)	Frequency
20.56–20.58	3
20.59–20.61	6
20.62–20.64	8
20.65–20.67	3

24.3   The Variance and Standard Deviation

Suppose we consider the test marks of two groups of three students. Suppose that the marks out of 10 for the first group are

4   7   and   10

while those of the second group are

7   7   and   7

It is easy to calculate the mean mark of each group: the first group has mean mark

$$\frac{4+7+10}{3} = \frac{21}{3} = 7$$

The second group has mean mark

$$\frac{7+7+7}{3} = \frac{21}{3} = 7$$

We see that both groups have the same mean mark even though the marks in the first group are widely spread, whilst the marks in the second group are all the same. If the teacher quotes just the mean mark of each group this gives no information about how widely the marks are spread. The **variance** and **standard deviation** are important and widely used statistical quantities which contain this information. Most calculators are pre-programmed to calculate these quantities. Once you understand the processes involved, check to see if your calculator can be used to find the variance and standard deviation of a set of data.

Suppose we have n values x_1, x_2, x_3 up to x_n. Their mean is $\bar{x}$ given by

$$\bar{x} = \frac{\sum_{i=1}^{n} x_i}{n}$$

The variance is found from the following formula.

KEY POINT

$$\text{variance} = \frac{\sum_{i=1}^{n} (x_i - \bar{x})^2}{n}$$

If you study this carefully you will see that to calculate the variance we must

- calculate the mean value $\bar{x}$

- subtract the mean from each value in turn, that is, find $x_i - \bar{x}$

- square each answer to get $(x_i - \bar{x})^2$

- add up all these squared quantities to get $\sum_{i=1}^{n}(x_i - \bar{x})^2$

- divide the result by n to get

$$\frac{\sum_{i=1}^{n}(x_i - \bar{x})^2}{n}; \text{ this is the variance.}$$

The standard deviation is found by taking the square root of the variance.

KEY POINT

$$\text{standard deviation} = \sqrt{\frac{\sum_{i=1}^{n}(x_i - \bar{x})^2}{n}}$$

Let us calculate the variance and standard deviation of each of the two sets of marks 4, 7, 10 and 7, 7, 7. We have already noted that the mean of each set is 7. Each stage of the calculation of the variance of the set 4, 7, 10 is shown in Table 24.5. The mean is subtracted from each number and the results are squared and then added. Note that when a negative

Table 24.5

x_i	$x_i - \bar{x}$	$(x_i - \bar{x})^2$
4	$4 - 7 = -3$	$(-3)^2 = 9$
7	$7 - 7 = 0$	$0^2 = 0$
10	$10 - 7 = 3$	$3^2 = 9$
Total		18

number is squared the result is positive. The squares are added to give 18. In this example, the number of values equals 3. So, taking $n = 3$,

$$\text{variance} = \frac{\sum_{i=1}^{n}(x_i - \bar{x})^2}{n} = \frac{18}{3} = 6$$

The standard deviation is the square root of the variance, that is, $\sqrt{6} = 2.449$. Similarly, calculation of the variance and standard deviation of the set 7, 7, 7 is shown in Table 24.6. Again, n equals 3. So

$$\text{variance} = \frac{\sum_{i=1}^{n}(x_i - \bar{x})^2}{n} = \frac{0}{3} = 0$$

The standard deviation is the square root of the variance and so it also equals zero. The fact that the standard deviation is zero reflects that there is no spread of values. By comparison it is easy to check that the standard deviation of the set 6, 7 and 8, which also has mean 7, is equal to 0.816.

Table 24:6

x_i	$x_i - \bar{x}$	$(x_i - \bar{x})^2$
7	$7 - 7 = 0$	$0^2 = 0$
7	$7 - 7 = 0$	$0^2 = 0$
7	$7 - 7 = 0$	$0^2 = 0$
Total		0

The fact that the set 4, 7 and 10 has standard deviation 2.449 shows that the values are more widely spread than in the set 6, 7 and 8.

Worked example

24.11 Find the variance and standard deviation of 10, 15.8, 19.2 and 8.7.

Solution First the mean is found:

$$\bar{x} = \frac{10 + 15.8 + 19.2 + 8.7}{4} = \frac{53.7}{4} = 13.425$$

The calculation to find the variance is given in Table 24.7:

Table 24.7

x_i	$x_i - \bar{x}$	$(x_i - \bar{x})^2$
10	$10 - 13.425 = -3.425$	$(-3.425)^2 = 11.731$
15.8	$15.8 - 13.425 = 2.375$	$2.375^2 = 5.641$
19.2	$19.2 - 13.425 = 5.775$	$5.775^2 = 33.351$
8.7	$8.7 - 13.425 = -4.725$	$(-4.725)^2 = 22.326$
Total		73.049

$$\text{variance} = \frac{\sum_{i=1}^{n}(x_i - \bar{x})^2}{n} = \frac{73.049}{4} = 18.262$$

The standard deviation is the square root of the variance: $\sqrt{18.262} = 4.273$.

When dealing with a grouped frequency distribution with N classes the formula for the variance becomes:

KEY POINT

$$\text{variance} = \frac{\sum_{i=1}^{N} f_i(x_i - \bar{x})^2}{\sum_{i=1}^{N} f_i}$$

As before the standard deviation is the square root of the variance.

Worked example

24.12   In a period of 30 consecutive days in July the temperature in °C was recorded as follows:

    18   19   20   23   24   24   21   18   17   16

    16   17   17   17   18   19   20   20   22   23

    24   24   25   23   21   21   20   19   19   18

(a) Produce a grouped frequency distribution showing data grouped from 16–17 °C, 18–19 °C and so on.

(b) Find the mean temperature of the grouped data.

(c) Find the standard deviation of the grouped data.

Solution   (a) The data is grouped using a tally chart as in Table 24.8.

Table 24.8

Temperature range (°C)	Tally	Frequency
16–17	ⅢⅠ	6
18–19	Ⅲ |||	8
20–21	Ⅲ ||	7
22–23	||||	4
24–25	Ⅲ	5
Total		30

(b) When calculating the mean temperature from the grouped data we do not know the actual temperatures. We do know the frequency of each class. The best we can do is use the midpoint of each class as an estimate of the values in that class. The class midpoints and the calculation to determine the mean are shown in Table 24.9. The mean is then

$$\bar{x} = \frac{\sum_{i=1}^{N} f_i \times x_i}{\sum_{i=1}^{N} f_i} = \frac{603}{30} = 20.1\,°C.$$

Table 24.9

Temperature range (°C)	Frequency f_i	Class midpoint x_i	$f_i \times x_i$
16–17	6	16.5	99.0
18–19	8	18.5	148.0
20–21	7	20.5	143.5
22–23	4	22.5	90.0
24–25	5	24.5	122.5
Total	30		603

(c) To find the variance and hence the standard deviation we must subtract the mean, 20.1, from each value, square and then add the results. Finally this sum is divided by 30. The complete calculation is shown in Table 24.10.

Table 24.10

Temperature range (°C)	Frequency f_i	Class midpoint x_i	$x_i - \bar{x}$	$(x_i - \bar{x})^2$	$f_i(x_i - \bar{x})^2$
16–17	6	16.5	−3.60	12.96	77.76
18–19	8	18.5	−1.60	2.56	20.48
20–21	7	20.5	0.40	0.16	1.12
22–23	4	22.5	2.40	5.76	23.04
24-25	5	24.5	4.40	19.36	96.80
Total	30				219.20

Finally the variance is given by

$$\text{variance} = \frac{\sum_{i=1}^{N} f_i(x_i - \bar{x})^2}{\sum_{i=1}^{N} f_i} = \frac{219.20}{30} = 7.307$$

The standard deviation is the square root of the variance, that is, $\sqrt{7.307} = 2.703$.

Self-assessment questions 24.3

1. Why is an average often an insufficient way of describing a set of values?
2. Describe the stages involved in calculating the variance and standard deviation of a set of values $x_1, x_2, \cdots, x_n$.

Exercise 24.3

1. Find the standard deviation of each of the following sets of numbers and comment upon your answers:

    (a) 20, 20, 20, 20 (b) 16, 17, 23, 24
    (c) 0, 20, 20, 40

2. Calculate the variance and standard deviation of the following set of numbers: 1, 2, 3, 3, 3, 4, 7, 8, 9, 10.

3. The examination results of two students, Jane and Tony, are shown in the table. Find the mean and standard deviation of each of them. Comment upon the results.

	Jane	Tony
Maths	50	29
Physics	42	41
Chemistry	69	60
French	34	48
Spanish	62	80

4. The marks of 50 students in a mathematics examination are given as

45	50	62	62	68	47	45	44	48
73	62	63	62	67	80	45	41	40
23	55	21	83	67	49	48	48	62
63	79	58	71	37	32	58	54	50
62	66	68	62	81	92	62	45	49
71	72	70	49	51				

    (a) Produce a grouped frequency distribution using classes 25–29, 30–34, 35–39 and so on.

    (b) The modal class is the class with the highest frequency. State the modal class.

    (c) Calculate the mean, variance and standard deviation of the grouped frequency distribution.

Test and assignment exercises 24

1. Calculate the mean, median and mode of the following numbers: 2.7, 2.8, 3.1, 3.1, 3.1, 3.4, 3.8, 4.1.

2. Calculate the mean of

    (a) the first ten whole numbers, 1, 2, $\cdots$, 10

    (b) the first 12 whole numbers, 1, 2, $\cdots$, 12

    (c) the first n whole numbers, 1, 2, $\cdots$, n.

3. A manufacturer of breakfast cereals uses two machines which pack 250 g packets of cereal automatically. In order to check the average weight, sample packets are taken and weighed to the nearest gram. The results of checking eight packets from each of the two machines are given in the table.

Machine								
1	250	250	251	257	253	250	268	259
2	249	251	250	251	252	258	254	250

(a) Find the mean weight of each sample of eight packets.

(b) Find the standard deviation of each sample.

(c) Comment upon the performance of the two machines.

4. What is the median of the five numbers 18, 28, 39, 42, 43? If 50 is added as the sixth number what would the median become?

5. Express the following sums in sigma notation:

(a) $y_1 + y_2 + y_3 + \cdots + y_7 + y_8$

(b) $y_1^2 + y_2^2 + y_3^2 + \cdots + y_7^2 + y_8^2$

(c) $(y_1 - \bar{y})^2 + (y_2 - \bar{y})^2 + \cdots + (y_7 - \bar{y})^2 + (y_8 - \bar{y})^2$

25 Probability

Objectives	This chapter
	• introduces theoretical and experimental probability and how to calculate them
	• explains the meaning of the term 'complementary events'
	• explains the meaning of the term 'independent events'

25.1 Introduction

Some events in life are impossible; other events are quite certain to happen. For example, it is impossible for a human being to live for 1000 years. We say that the probability of a human being living for 1000 years is zero. On the other hand it is quite certain that all of us will die someday. We say that the probability that all of us will die is 1. Using the letter P to stand for probability we can write

P(a human will live for 1000 years) $= 0$

and

P(all of us will die someday) $= 1$

Many other events in life are neither impossible nor certain. These have varying degrees of likelihood, or chance. For example, it is quite unlikely but not impossible that the British weather throughout the summer of 1998 will be dominated by snowfall. It is quite likely but not certain that the life expectancy of the UK population will continue to rise in the foreseeable future. Events such as these can be assigned probabilities varying from 0 up to 1. Those having probabilities close to 1 are quite likely to happen. Those having probabilities close to 0 are almost impossible. Several events and their probabilities are shown on the probability scale in Figure 25.1. It is important to note that no probability can lie outside the range 0 to 1.

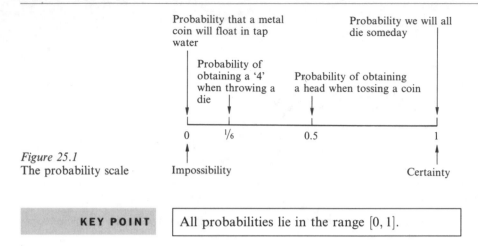

Figure 25.1
The probability scale

KEY POINT All probabilities lie in the range [0, 1].

Complementary events

Consider the following situation. A light bulb is tested. Clearly it either works or it does not work. Here we have two events: the first is that the light bulb works and the second is that the light bulb does not work. When the bulb is tested, one or other of these events must occur. Furthermore, each event excludes the other. In such a situation we say the two events are **complementary**. In general, two events are complementary if one of them must happen and, when it does, the other event cannot. The sum of the probabilities of the two complementary events must always equal 1. This is known as **total probability**. We shall see how this result can be used to calculate probabilities shortly.

Self-assessment questions 25.1

1. All probabilities must lie in a certain range. What is this range?
2. Explain the meaning of the term 'complementary events'.

| 25.2 | **Calculating theoretical probabilities** |

Sometimes we have sufficient information about a set of circumstances to calculate the probability of an event occurring. For example, suppose we roll a die and ask what is the probability of obtaining a score of '5'. If the die were fair, or **unbiased**, the chance of getting a '5' is the same as the chance of getting any other score. You would expect to get a '5' one time in every six. That is,

the probability of throwing a '5' is $\frac{1}{6}$ or 0.167

In other words there is a one in six chance of scoring a '5'. The fact that the probability is closer to 0 than to 1 means that it is quite unlikely that a '5' will be thrown although it is not impossible. Such a probability is known as a **theoretical probability** and, when all events are equally likely, it is calculated from the following formula.

KEY POINT

$$P\left(\begin{array}{c}\text{obtaining our} \\ \text{chosen event}\end{array}\right) = \frac{\text{number of ways the chosen event can occur}}{\text{total number of possibilities}}$$

For example, suppose we ask what is the probability of obtaining a score more than 4. A score more than 4 can occur in two ways, by scoring a '5' or a '6'. If the die is fair, all possible events are equally likely. The total number of possibilities is 6, therefore,

$$P(\text{score more than 4}) = \frac{2}{6} = \frac{1}{3}$$

Worked examples

25.1   A fair die is thrown. What is the probability of obtaining an even score?

Solution   The chosen event is throwing an even score, that is, a '2', '4' or '6'. There are therefore three ways that this chosen event can occur out of a total of six, equally likely, ways. So

$$P(\text{even score}) = \frac{3}{6} = \frac{1}{2}$$

25.2   A fair coin is tossed. What is the probability that it will land with its head uppermost?

Solution   There are two equally likely ways the coin can land: head uppermost or tail uppermost. The chosen event, that the coin lands with its head uppermost, is just one of these ways. The chance of getting a head is the same as the chance of getting a tail. Therefore,

$$P(\text{head}) = \frac{1}{2}$$

Note from the previous example that $P(\text{tail}) = \frac{1}{2}$ and that also

$$P(\text{tail}) + P(\text{head}) = 1$$

Note that the two events, getting a head and getting a tail, are complementary because one of them must happen and either one excludes the other. Therefore the sum of the two probabilities, the total probability, equals 1.

Worked example

25.3   Two coins are tossed.

(a) Write down all the possible outcomes.

(b) What is the probability of obtaining two tails?

Solution   (a) Letting H stand for head and T for tail, the possible outcomes are:

$$H, H \qquad H, T \qquad T, H \qquad T, T$$

There are four possible outcomes, each one equally likely to occur.

(b) Obtaining two tails is just one of the four possible outcomes. Therefore, the probability of obtaining two tails is $\frac{1}{4}$.

Exercise 25.2

1. A die is thrown. Find

   (a) the probability of obtaining a score less than 6,
   (b) the probability of obtaining a score more than 6,
   (c) the probability of obtaining an even score less than 5,
   (d) the probability of obtaining an even score less than 2.

2. There are four Aces in a pack of 52 playing cards. What is the probability that a card selected at random is not an Ace?

3. A drawer contains six red socks, six black socks and eight blue socks. Find the probability that a sock selected at random from the drawer is

   (a) black (b) red (c) red or blue

4. Two dice are thrown together and their scores are added together. By considering all the possible outcomes, find the probability that the total score will be

   (a) 12 (b) 0 (c) 1 (d) 2 (e) more than 5

5. A basket contains 87 good apples and three bad ones. What is the probability that an apple chosen at random is bad?

25.3   **Calculating experimental probabilities**

In some circumstances we do not have sufficient information to calculate a theoretical probability. We know that if a coin is unbiased the probability of obtaining a head is $\frac{1}{2}$. But suppose the coin is biased so that it is more likely to land with its tail uppermost. We can experiment by tossing the coin a large number of times and counting the number of tails obtained. Suppose we toss the coin 100 times and obtain 65 tails. We can then estimate the probability of obtaining a tail as $\frac{65}{100}$ or 0.65. Such a probability is known as an **experimental probability** and is only accurate if a very large number of experiments are performed. Generally, we can calculate an experimental probability from the following formula.

KEY POINT

$$P\binom{\text{chosen event}}{\text{occurs}} = \frac{\text{number of ways the chosen event occurs}}{\text{total number of times the experiment is repeated}}$$

Worked examples

25.4   A biased die is thrown 1000 times and a score of '6' is obtained on 200 occasions. If the die is now thrown again what is the probability of obtaining a score of '6'?

Solution   Using the formula for the experimental probability we find

$$P(\text{throwing a six}) = \frac{200}{1000} = 0.2$$

If the die were unbiased the theoretical probability of throwing a six would be $\frac{1}{6} = 0.167$, so the die has been biased in favour of throwing a '6'.

25.5   A manufacturer produces microwave ovens. It is known from experience that the probability that a microwave oven is of an acceptable standard is 0.92. Find the probability that an oven selected at random is not of an acceptable standard.

Solution   When an oven is tested either it is of an acceptable standard or it is not. An oven cannot be both acceptable and unacceptable. The two events, that the oven is acceptable or that the oven is unacceptable, are therefore complementary. Recall that the sum of the probabilities of complementary events is 1 and so

$$P(\text{oven is not acceptable}) = 1 - 0.92 = 0.08$$

Self-assessment questions 25.3

1. In what circumstances is it appropriate to use an experimental probability?
2. A series of experiments is performed three times. In the first of the series an experiment is carried out 100 times. In the second it is carried out 1000 times and in the third 10000 times. Which of the series of experiments is likely to lead to the best estimate of probability? Why?

Exercise 25.3

1. A new component is fitted to a washing machine. In a sample of 150 machines tested, seven failed to function correctly. Calculate the probability that a machine fitted with the new component (a) works correctly, (b) does not work correctly.

2. In a sample containing 5000 nails manufactured by a company, 5% are too short or too long. A nail is picked at random from the production line; estimate the probability that it is of the right length.

3. The probability that the post office delivers first class mail on the following working day after posting is 0.96. If 93500 first class letters are posted on Wednesday, how many are likely to be delivered on Thursday?

25.4 Independent events

Two events are **independent** if the occurrence of either one in no way affects the occurrence of the other. For example, if an unbiased die is thrown twice the score on the second throw is in no way affected by the score on the first. The two scores are independent. The **multiplication law** for independent events states the following:

KEY POINT

> If events A and B are independent, then the probability of obtaining A and B is given by
> $$P(A \text{ and } B) = P(A) \times P(B)$$

Worked example

25.6 A die is thrown and a coin is tossed. What is the probability of obtaining a six and a head?

Solution These events are independent since the score on the die in no way affects the result of tossing the coin, and vice versa. Therefore

$$P(\text{throwing a '6' and tossing a head}) = P(\text{throwing a '6'})$$
$$\times P(\text{tossing a head})$$
$$= \frac{1}{6} \times \frac{1}{2}$$
$$= \frac{1}{12}$$

When several events are independent of each other the multiplication law becomes

$$P(A \text{ and } B \text{ and } C \text{ and } D \ldots) = P(A) \times P(B) \times P(C) \times P(D) \ldots$$

Worked example

25.7   A coin is tossed three times. What is the probability of obtaining three heads?

Solution   The three tosses are all independent events since the result of any one has no effect on the others. Therefore

$$P(3 \text{ heads}) = \frac{1}{2} \times \frac{1}{2} \times \frac{1}{2} = \frac{1}{8}$$

Self-assessment questions 25.4

1. Explain what is meant by saying two events are independent.
2. Suppose we have two packs each of 52 playing cards. A card is selected from each pack. Event A is that the card from the first pack is the Ace of Spades. Event B is that the card from the second pack is a Spade. Are these two events dependent or independent?
3. From a single pack of cards, two are removed. The first is examined. Suppose event A is that the first card is the Ace of Spades. The second card is examined. Event B is that the second card is a Spade. Are the two events dependent or independent?

Exercise 25.4

1. A die is thrown and a coin is tossed. What is the probability of getting an even score on the die and a tail?

2. Suppose you have two packs each of 52 playing cards. A card is drawn from the first and a card is drawn from the second. What is the probability that both cards are the Ace of Spades?

3. A coin is tossed eight times. What is the probability of obtaining eight tails?

4. A die is thrown four times. What is the probability of obtaining four 1s?

5. Suppose that there is an equal chance of a mother giving birth to a boy or a girl.

   (a) If one child is born find the probability that it is a boy.

   (b) If two children are born find the probability that they are both boys, assuming that the sex of neither one can influence the sex of the other.

Test and assignment exercises 25

1. The following numbers are probabilities of a certain event happening. One of them is an error. Which one?

   (a) 0.5 (b) $\frac{3}{4}$ (c) 0.001 (d) $\frac{13}{4}$

2. Which of the following cannot be a probability?

   (a) 0.125 (b) −0.2 (c) 1 (d) 0

3. Events A, B and C are defined as follows. In each case state the complementary event.

   A: the lifespan of a light bulb is greater than 1000 hours
   B: it will rain on Christmas Day 1999
   C: your car will be stolen sometime in the next 100 days

4. The probability of a component manufactured in a factory being defective is 0.01. If three components are selected at random what is the probability that they all work as required?

5. A parcel delivery company guarantees 'next-day' delivery for 99% of its parcels. Out of a sample of 1800 deliveries 15 were delivered later than the following day. Is the company living up to its promises?

6. At a certain bus stop the probability of a bus arriving late is $\frac{3}{20}$. The probability of it arriving on time is $\frac{4}{5}$. Find

   (a) the probability that the bus arrives early,

   (b) the probability that the bus does not arrive late.

7. Suppose you have two packs each of 52 playing cards. A card is drawn from the first and a card is drawn from the second. What is the probability that

   (a) both cards are diamonds?

   (b) both cards are black?

   (c) both cards are Kings?

8. Find the probability of obtaining an odd number when throwing a fair die once.

9. A bag contains eight red beads, four white beads and five blue beads. A bead is drawn at random. Determine the probability that it is

   (a) red (b) white (c) black (d) blue
   (e) red or white (f) not blue.

10. Two cards are selected from a pack of 52. Find the probability that they are both Aces if the first card is (a) replaced and (b) not replaced.

Solutions to exercises

Solutions to Chapter 1

Exercises 1.1
1. (a) $\{x : x \in \mathbb{R}, 2 \le x \le 6\}$

The interval $[2,6]$

(b) $\{x : x \in \mathbb{R}, 6 < x \le 8\}$

The interval $(6,8]$

(c) $\{x : x \in \mathbb{R}, -2 < x < 0\}$

The interval $(-2,0)$

(d) $\{x : x \in \mathbb{R}, -3 \le x < -1.5\}$

The interval $[-3,-1.5)$

2. (a) 3 is a member of $\mathbb{Z}$, $\mathbb{Z}^+$ and $\mathbb{N}$; (b) -3 is a member of $\mathbb{Z}$ and $\mathbb{Z}^-$; (c) 3.2 does not belong to any of the sets; (d) 11 is a member of $\mathbb{Z}$, $\mathbb{Z}^+$ and $\mathbb{N}$; (e) 0.01 does not belong to any of these sets.
3. (a) T; (b) T; (c) T; (d) T; (e) F; (f) T; (g) F; (h) T.

Exercises 1.2
1. (a) 6960; (b) 70.4; (c) 0.0123; (d) 0.0110; (e) 45.6; (f) 2350.

2. (a) 66.00, (b) 66.0; (c) 66; (d) 70; (e) 66.00; (f) 66.0.
3. (a) 10; (b) 10.0.
4. (a) 65.456; (b) 65.46; (c) 65.5; (d) 65.456; (e) 65.46; (f) 65.5; (g) 65; (h) 70.

Exercises 1.3
1. (a) $6 - 2 \times 2 = 6 - 4 = 2$;
   (b) $(6 - 2) \times 2 = 4 \times 2 = 8$;
   (c) $6 \div 2 - 2 = 3 - 2 = 1$;
   (d) $(6 \div 2) - 2 = 3 - 2 = 1$;
   (e) $6 - 2 + 3 \times 2 = 6 - 2 + 6 = 10$;
   (f) $6 - (2 + 3) \times 2 = 6 - 5 \times 2$
   $= 6 - 10 = -4$;
   (g) $(6 - 2) + 3 \times 2 = 4 + 3 \times 2 = 4 + 6 = 10$;
   (h) $\frac{16}{-2} = -8$;
   (i) $\frac{-24}{-3} = 8$;
   (j) $(-6) \times (-2) = 12$;
   (k) $(-2)(-3)(-4) = -24$.
2. (a) Sum $= 3 + 6 = 9$,
   Product $= 3 \times 6 = 18$;
   (b) Sum $= 17$, Product $= 70$;
   (c) Sum $= 2 + 3 + 6 = 11$,
   Product $= 2 \times 3 \times 6 = 36$.
3. (a) Difference $= 18 - 9 = 9$,
   Quotient $= \frac{18}{9} = 2$;
   (b) Difference $= 20 - 5 = 15$,
   Quotient $= \frac{20}{5} = 4$;
   (c) Difference $= 100 - 20 = 80$,
   Quotient $= \frac{100}{20} = 5$.
4. (a) $6 \times (12 - 3) + 1 = 6 \times 9 + 1$
   $= 54 + 1 = 55$;

(b) $6 \times 12 - (3 + 1) = 72 - 4 = 68$;

(c) $6 \times (12 - 3 + 1) = 6 \times 10 = 60$;

(d) $5 \times (4 - 3) + 2 = 5 \times 1 + 2 = 5 + 2 = 7$;

(e) $5 \times 4 - (3 + 2) = 20 - 5 = 15$

or $5 \times (4 - 3 + 2) = 5 \times 3 = 15$;

(f) $5 \times (4 - (3 + 2)) = 5 \times (4 - 5)$
$= 5 \times (-1) = -5$.

Exercises 1.4

1. (a) 2×13; (b) $2 \times 2 \times 5 \times 5$; (c) $3 \times 3 \times 3$;
   (d) 71; (e) $2 \times 2 \times 2 \times 2 \times 2 \times 2$; (f) 3×29;
   (g) 19×23; (h) 29×31.

Exercises 1.5

1. (a) $12 = 2 \times 2 \times 3$ $15 = 3 \times 5$ $21 = 3 \times 7$
   Hence h.c.f. $= 3$.

   (b) $16 = 2 \times 2 \times 2 \times 2$ $24 = 2 \times 2 \times 2 \times 3$
   $40 = 2 \times 2 \times 2 \times 5$
   So h.c.f. $= 2 \times 2 \times 2 = 8$.

   (c) $28 = 2 \times 2 \times 7$ $70 = 2 \times 5 \times 7$
   $120 = 2 \times 2 \times 2 \times 3 \times 5$
   $160 = 2 \times 2 \times 2 \times 2 \times 2 \times 5$

So h.c.f. $= 2$.

(d) $35 = 5 \times 7$ $38 = 2 \times 19$
$42 = 2 \times 3 \times 7$
So h.c.f. $= 1$.

(e) $96 = 2 \times 2 \times 2 \times 2 \times 2 \times 3$
$120 = 2 \times 2 \times 2 \times 3 \times 5$
$144 = 2 \times 2 \times 2 \times 2 \times 3 \times 3$
So h.c.f. $= 2 \times 2 \times 2 \times 3 = 24$.

2. (a) 5 $6 = 2 \times 3$ $8 = 2 \times 2 \times 2$
   So l.c.m. $= 2 \times 2 \times 2 \times 3 \times 5 = 120$.

   (b) $20 = 2 \times 2 \times 5$ $30 = 2 \times 3 \times 5$
   So l.c.m. $= 2 \times 2 \times 3 \times 5 = 60$.

   (c) 7 $9 = 3 \times 3$ $12 = 2 \times 2 \times 3$
   So l.c.m. $= 2 \times 2 \times 3 \times 3 \times 7 = 252$.

   (d) $100 = 2 \times 2 \times 5 \times 5$
   $150 = 2 \times 3 \times 5 \times 5$
   $235 = 5 \times 47$
   So l.c.m. $=$
   $2 \times 2 \times 3 \times 5 \times 5 \times 47 = 14100$.

   (e) $96 = 2 \times 2 \times 2 \times 2 \times 2 \times 3$
   $120 = 2 \times 2 \times 2 \times 3 \times 5$
   $144 = 2 \times 2 \times 2 \times 2 \times 3 \times 3$
   So l.c.m. $=$
   $2 \times 2 \times 2 \times 2 \times 2 \times 3 \times 3 \times 5 = 1440$.

Solutions to Chapter 2

Exercises 2.1

1. (a) $\dfrac{5}{3} = \dfrac{3 + 2}{3} = 1\dfrac{2}{3}$

   (b) $\dfrac{9}{4} = \dfrac{8 + 1}{4} = 2\dfrac{1}{4}$

   (c) $\dfrac{7}{5} = \dfrac{5 + 2}{5} = 1\dfrac{2}{5}$

   (d) $-\dfrac{20}{11} = -\dfrac{11 + 9}{11} = -1\dfrac{9}{11}$

   (e) $\dfrac{11}{5} = \dfrac{10 + 1}{5} = 2\dfrac{1}{5}$

2. (a) $2\dfrac{1}{2} = 2 + \dfrac{1}{2} = \dfrac{4}{2} + \dfrac{1}{2} = \dfrac{5}{2}$

   (b) $3\dfrac{2}{3} = 3 + \dfrac{2}{3} = \dfrac{9}{3} + \dfrac{2}{3} = \dfrac{11}{3}$

(c) $10\dfrac{1}{4} = 10 + \dfrac{1}{4} = \dfrac{40}{4} + \dfrac{1}{4} = \dfrac{41}{4}$

(d) $5\dfrac{2}{7} = 5 + \dfrac{2}{7} = \dfrac{35}{7} + \dfrac{2}{7} = \dfrac{37}{7}$

(e) $6\dfrac{2}{9} = 6 + \dfrac{2}{9} = \dfrac{54}{9} + \dfrac{2}{9} = \dfrac{56}{9}$

Exercises 2.2

1. (a) $\dfrac{18}{27} = \dfrac{2}{3}$ (b) $\dfrac{12}{20} = \dfrac{3}{5}$ (c) $\dfrac{15}{45} = \dfrac{1}{3}$

   (d) $\dfrac{25}{80} = \dfrac{5}{16}$ (e) $\dfrac{15}{60} = \dfrac{1}{4}$ (f) $\dfrac{90}{200} = \dfrac{9}{20}$

(g) $\dfrac{15}{20}=\dfrac{3}{4}$ (h) $\dfrac{2}{18}=\dfrac{1}{9}$ (i) $\dfrac{16}{24}=\dfrac{2}{3}$

(j) $\dfrac{30}{65}=\dfrac{6}{13}$ (k) $\dfrac{12}{21}=\dfrac{4}{7}$ (l) $\dfrac{100}{45}=\dfrac{20}{9}$

Exercises 2.3

1. (a) $\dfrac{1}{4}+\dfrac{2}{3}=\dfrac{3}{12}+\dfrac{8}{12}=\dfrac{11}{12}$

(b) $\dfrac{3}{5}+\dfrac{5}{3}=\dfrac{9}{15}+\dfrac{25}{15}=\dfrac{34}{15}$

(c) $\dfrac{12}{14}-\dfrac{2}{7}=\dfrac{6}{7}-\dfrac{2}{7}=\dfrac{4}{7}$

(d) $\dfrac{3}{7}-\dfrac{1}{2}+\dfrac{2}{21}=\dfrac{18}{42}-\dfrac{21}{42}+\dfrac{4}{42}=\dfrac{1}{42}$

(e) $1\dfrac{1}{2}+\dfrac{4}{9}=\dfrac{3}{2}+\dfrac{4}{9}=\dfrac{27}{18}+\dfrac{8}{18}=\dfrac{35}{18}$

(f) $2\dfrac{1}{4}-1\dfrac{1}{3}+\dfrac{1}{2}=\dfrac{9}{4}-\dfrac{4}{3}+\dfrac{1}{2}$

$=\dfrac{27}{12}-\dfrac{16}{12}+\dfrac{6}{12}=\dfrac{17}{12}$

(g) $\dfrac{10}{15}-1\dfrac{2}{5}+\dfrac{8}{3}=\dfrac{10}{15}-\dfrac{7}{5}+\dfrac{8}{3}$

$=\dfrac{10}{15}-\dfrac{21}{15}+\dfrac{40}{15}=\dfrac{29}{15}$

(h) $\dfrac{9}{10}-\dfrac{7}{16}+\dfrac{1}{2}-\dfrac{2}{5}=\dfrac{72}{80}-\dfrac{35}{80}+\dfrac{40}{80}-\dfrac{32}{80}$

$=\dfrac{45}{80}=\dfrac{9}{16}$

Exercises 2.4

1. (a) $\dfrac{2}{3}\times\dfrac{6}{7}=\dfrac{2}{1}\times\dfrac{2}{7}=\dfrac{4}{7}$

(b) $\dfrac{8}{15}\times\dfrac{25}{32}=\dfrac{1}{15}\times\dfrac{25}{4}=\dfrac{1}{3}\times\dfrac{5}{4}=\dfrac{5}{12}$

(c) $\dfrac{1}{4}\times\dfrac{8}{9}=\dfrac{1}{1}\times\dfrac{2}{9}=\dfrac{2}{9}$

(d) $\dfrac{16}{17}\times\dfrac{34}{48}=\dfrac{1}{17}\times\dfrac{34}{3}=\dfrac{1}{1}\times\dfrac{2}{3}=\dfrac{2}{3}$

(e) $2\times\dfrac{3}{5}\times\dfrac{5}{12}=\dfrac{3}{5}\times\dfrac{5}{6}=\dfrac{3}{1}\times\dfrac{1}{6}=\dfrac{1}{2}$

(f) $2\dfrac{1}{3}\times1\dfrac{1}{4}=\dfrac{7}{3}\times\dfrac{5}{4}=\dfrac{35}{12}$

(g) $1\dfrac{3}{4}\times2\dfrac{1}{2}=\dfrac{7}{4}\times\dfrac{5}{2}=\dfrac{35}{8}$

(h) $\dfrac{3}{4}\times1\dfrac{1}{2}\times3\dfrac{1}{2}=\dfrac{3}{4}\times\dfrac{3}{2}\times\dfrac{7}{2}=\dfrac{63}{16}$

2. (a) $\dfrac{2}{3}\times\dfrac{3}{4}=\dfrac{1}{2}$

(b) $\dfrac{4}{7}\times\dfrac{21}{30}=\dfrac{6}{15}=\dfrac{2}{5}$

(c) $\dfrac{9}{10}\times80=72$

(d) $\dfrac{6}{7}\times42=36$

3. Yes, because $\dfrac{3}{4}\times\dfrac{12}{15}=\dfrac{12}{15}\times\dfrac{3}{4}$

Exercises 2.5

1. (a) $\dfrac{3}{4}\div\dfrac{1}{8}=\dfrac{3}{4}\times\dfrac{8}{1}=\dfrac{3}{1}\times\dfrac{2}{1}=6$

(b) $\dfrac{8}{9}\div\dfrac{4}{3}=\dfrac{8}{9}\times\dfrac{3}{4}=\dfrac{2}{9}\times\dfrac{3}{1}=\dfrac{2}{3}$

(c) $-\dfrac{2}{7}\div\dfrac{4}{21}=-\dfrac{2}{7}\times\dfrac{21}{4}=-\dfrac{2}{1}\times\dfrac{3}{4}=-\dfrac{3}{2}$

(d) $\dfrac{9}{4}\div1\dfrac{1}{2}=\dfrac{9}{4}\div\dfrac{3}{2}=\dfrac{9}{4}\times\dfrac{2}{3}=\dfrac{3}{4}\times\dfrac{2}{1}=\dfrac{3}{2}$

(e) $\dfrac{5}{6}\div\dfrac{5}{12}=\dfrac{5}{6}\times\dfrac{12}{5}=\dfrac{1}{6}\times\dfrac{12}{1}=2$

(f) $\dfrac{99}{100}\div1\dfrac{4}{5}=\dfrac{99}{100}\div\dfrac{9}{5}=\dfrac{99}{100}\times\dfrac{5}{9}$

$=\dfrac{11}{100}\times\dfrac{5}{1}=\dfrac{11}{20}$

(g) $3\dfrac{1}{4}\div1\dfrac{1}{8}=\dfrac{13}{4}\div\dfrac{9}{8}=\dfrac{13}{4}\times\dfrac{8}{9}$

$=\dfrac{13}{1}\times\dfrac{2}{9}=\dfrac{26}{9}$

(h) $\left(2\dfrac{1}{4}\div\dfrac{3}{4}\right)\times2=\left(\dfrac{9}{4}\times\dfrac{4}{3}\right)\times2$

$=\left(\dfrac{3}{4}\times4\right)\times2=3\times2=6$

(i) $2\dfrac{1}{4} \div \left(\dfrac{3}{4} \times 2\right) = \dfrac{9}{4} \div \left(\dfrac{3}{4} \times \dfrac{2}{1}\right)$

$= \dfrac{9}{4} \div \dfrac{3}{2} = \dfrac{9}{4} \times \dfrac{2}{3} = \dfrac{3}{4} \times \dfrac{2}{1} = \dfrac{3}{2}$

(j) $6\dfrac{1}{4} \div 2\dfrac{1}{2} + 5 = \dfrac{25}{4} \div \dfrac{5}{2} + 5$

$= \dfrac{25}{4} \times \dfrac{2}{5} + 5 = \dfrac{5}{2} + 5 = \dfrac{15}{2}$

(k) $6\dfrac{1}{4} \div \left(2\dfrac{1}{2} + 5\right) = \dfrac{25}{4} \div \left(\dfrac{5}{2} + 5\right)$

$= \dfrac{25}{4} \div \dfrac{15}{2} = \dfrac{25}{4} \times \dfrac{2}{15} = \dfrac{5}{4} \times \dfrac{2}{3} = \dfrac{5}{6}$

Solutions to Chapter 3

Exercises 3.1

1. 23% of $124 = \dfrac{23}{100} \times 124 = 28.52$

2. (a) $\dfrac{9}{11} = \dfrac{9}{11} \times 100\% = \dfrac{900}{11}\% = 81.82\%$

   (b) $\dfrac{15}{20} = \dfrac{15}{20} \times 100\% = 75\%$

   (c) $\dfrac{9}{10} = \dfrac{9}{10} \times 100\% = 90\%$

   (d) $\dfrac{45}{50} = \dfrac{45}{50} \times 100\% = 90\%$

   (e) $\dfrac{75}{90} = \dfrac{75}{90} \times 100\% = 83.33\%$

3. $\dfrac{13}{12} = \dfrac{13}{12} \times 100\% = 108.33\%$

4. 217% of $500 = \dfrac{217}{100} \times 500 = 1085$

5. New weekly wage is 106% of £400,

$$106\% \text{ of } 400 = \dfrac{106}{100} \times 400 = 424$$

The new weekly wage is £424.

6. 17% of $1200 = \dfrac{17}{100} \times 1200 = 204$

The debt is decreased by £204 to £1200 − 204 = £996.

7. (a) $50\% = \dfrac{50}{100} = 0.5$

   (b) $36\% = \dfrac{36}{100} = 0.36$

   (c) $75\% = \dfrac{75}{100} = 0.75$

   (d) $100\% = \dfrac{100}{100} = 1$

   (e) $12.5\% = \dfrac{12.5}{100} = 0.125$

Exercises 3.2

1. $8 + 1 + 3 = 12$. The first number is $\dfrac{8}{12}$ of 180, that is, 120; the second number is $\dfrac{1}{12}$ of 180, that is, 15; and the third number is $\dfrac{3}{12}$ of 180, that is, 45. Hence 180 is divided into 120, 15 and 45.

2. $1 + 1 + 3 = 5$. We calculate $\dfrac{1}{5}$ of 930 to be 186 and $\dfrac{3}{5}$ of 930 to be 558. The length is divided into 186 cm, 186 cm and 558 cm.

3. $2 + 3 + 4 = 9$. The first piece is $\dfrac{2}{9}$ of 6 m, that is, 1.33 m; the second piece is $\dfrac{3}{9}$ of 6 m, that is, 2 m; the third piece is $\dfrac{4}{9}$ of 6 m, that is, 2.67 m.

4. $1 + 2 + 3 + 4 = 10$: $\dfrac{1}{10}$ of $1200 = 120$; $\dfrac{2}{10}$ of $1200 = 240$; $\dfrac{3}{10}$ of $1200 = 360$; $\dfrac{4}{10}$ of $1200 = 480$. The number 1200 is divided into 120, 240, 360 and 480.

5. $2\dfrac{3}{4} : 1\dfrac{1}{2} : 2\dfrac{1}{4} = \dfrac{11}{4} : \dfrac{3}{2} : \dfrac{9}{4} = 11 : 6 : 9$

Now, $11 + 6 + 9 = 26$, so

$$\dfrac{11}{26} \times 2600 = 1100 \qquad \dfrac{6}{26} \times 2600 = 600$$

$$\dfrac{9}{26} \times 2600 = 900$$

Alan receives £1100, Bill receives £600 and Claire receives £900.

Solutions to Chapter 4

Exercises 4.2

1. $2^4 - 16$; $\left(\frac{1}{2}\right)^2 = \frac{1}{4}$; $1^8 = 1$; $3^5 = 243$; $0^3 = 0$.
2. $10^4 = 10000$; $10^5 = 100000$; $10^6 = 1000000$.
3. $11^4 = 14641$; $16^8 = 4294967296$;
   $39^4 = 2313441$; $1.5^7 = 17.0859375$.
4. (a) $a^4 b^2 c = a \times a \times a \times a \times b \times b \times c$;
   (b) $xy^2 z^4 = x \times y \times y \times z \times z \times z \times z$.
5. (a) $x^4 y^2$; (b) $x^2 y^2 z^3$; (c) $x^2 y^2 z^2$; (d) $a^2 b^2 c^2$.
6. (a) $7^4 = 2401$; (b) $7^5 = 16807$;
   (c) $7^4 \times 7^5 = 40353607$;
   (d) $7^9 = 40353607$; (e) $8^3 = 512$;
   (f) $8^7 = 2097152$;
   (g) $8^3 \times 8^7 = 1073741824$;
   (h) $8^{10} = 1073741824$.
   The rule states that $a^m \times a^n = a^{m+n}$; the
   powers are added.
7. $(-3)^3 = -27$; $(-2)^2 = 4$; $(-1)^7 = -1$;
   $(-1)^4 = 1$.
8. -4492.125; 324; -0.03125.

Exercises 4.3

1. 60.
2. 69.
3. (a) 314.2 cm^2; (b) 28.27 cm^2; (c) 0.126 cm^2.
4. $3x^2 = 3 \times 4^2 = 3 \times 16 = 48$;
   $(3x)^2 = 12^2 = 144$.
5. $5x^2 = 5(-2)^2 = 20$; $(5x)^2 = (-10)^2 = 100$.
6. (a) 33.95; (b) 23.5225; (c) 26.75;
   (d) 109.234125.
7. (a) $a + b + c = 25.5$; (b) $ab = 46.08$;
   (c) $bc = 32.76$; (d) $abc = 419.328$.
8. $C = \frac{5}{9}(100 - 32) = \frac{5}{9}(68) = 37.78$.
9. (a) $x^2 = 49$; (b) $-x^2 = -49$;
   (c) $(-x)^2 = (-7)^2 = 49$.
10. $x^2 - 7x + 2 = (-9)^2 - 7(-9) + 2$
    $= 81 + 63 + 2 = 146$.
11. $2x^2 + 3x - 11 = 2(-3)^2 + 3(-3) - 11$
    $= 18 - 9 - 11 = -2$.
12. $-x^2 + 3x - 5 = -(-1)^2 + 3(-1) - 5$
    $= -1 - 3 - 5 = -9$.
13. 0.
14. (a) 49; (b) 43; (c) 1; (d) 5.

Solutions to Chapter 5

Exercises 5.1

1. (a) $5^7 \times 5^{13} = 5^{20}$; (b) $9^8 \times 9^5 = 9^{13}$;
   (c) $11^2 \times 11^3 \times 11^4 = 11^9$.
2. (a) $15^3/15^2 = 15^1 = 15$; (b) $4^{18}/4^9 = 4^9$;
   (c) $5^{20}/5^{19} = 5^1 = 5$.
3. (a) a^{10}; (b) a^9; (c) b^{22}.
4. (a) x^{15}; (b) y^{21}; (c) $15x^{15}$.
5. $19^8 \times 17^8$ cannot be simplified using the
   laws of indices because the two bases are
   not the same.
6. (a) $(7^3)^2 = 7^{3\times2} = 7^6$; (b) $(4^2)^8 = 4^{16}$;
   (c) $(7^9)^2 = 7^{18}$.
7. $1/(5^3)^8 = 1/5^{24}$.
8. (a) $x^5 y^5$; (b) $6x^6$; (c) $a^3 b^3 c^3$; (d) $2c^2 d$.

Exercises 5.2

1. (a) a^{11}; (b) x^{-16}; (c) x^{-18}; (d) 4^{-6}.
2. (a) 0.001; (b) 0.0001; (c) 0.00001.
3. $4^{-8}/4^{-6} = 4^{-2} = 1/4^2 = \frac{1}{16}$;
   $3^{-5}/3^{-8} = 3^3 = 27$.

Exercises 5.3

1. (a) $64^{1/3} = \sqrt[3]{64} = 4$ since $4^3 = 64$;
   (b) $144^{1/2} = \sqrt{144} = \pm12$;
   (c) $16^{-1/4} = 1/16^{1/4} = 1/\sqrt[4]{16} = \pm\frac{1}{2}$;

(d) $25^{-1/2} = 1/25^{1/2} = 1/\sqrt{25} = \pm\frac{1}{5}$;

(e) $1/32^{-1/5} = 32^{1/5} = \sqrt[5]{32} = 2$ since
$2^5 = 32$.

2. (a) $(3^{-1/2})^4 = 3^{-2} = 1/3^2 = \frac{1}{9}$;

(b) $(8^{1/3})^{-1} = 8^{-1/3} = 1/8^{1/3} = 1/\sqrt[3]{8} = \frac{1}{2}$.

Exercises 5.4

1. (a) $45 = 4.5 \times 10^1$; (b) $45000 = 4.5 \times 10^4$;

(c) $-450 = -4.5 \times 10^2$;

(d) $90000000 = 9.0 \times 10^7$;

(e) $0.15 = 1.5 \times 10^{-1}$;

(f) $0.00036 = 3.6 \times 10^{-4}$; (g) 3.5 is already in standard form;

(h) $-13.2 = -1.32 \times 10^1$;

(i) $1000000 = 1 \times 10^6$;

(j) $0.0975 = 9.75 \times 10^{-2}$;

(k) $45.34 = 4.534 \times 10^1$.

2. (a) $3.75 \times 10^2 = 375$; (b) $3.97 \times 10^1 = 39.7$;

(c) $1.875 \times 10^{-1} = 0.1875$;

(d) $-8.75 \times 10^{-3} = -0.00875$.

Solutions to Chapter 6

Exercises 6.1

1. (a) $-5p + 19q$; (b) $-5r - 13s + z$; (c) not possible to simplify;

(d) $8x^2 + 3y^2 - 2y$; (e) $4x^2 - x + 9$.

2. (a) $-12y + 8p + 9q$; (b) $21x^2 - 11x^3 + y^3$.

Exercises 6.2

1. (a) $-21z^2$; (b) $-4z$.

2. (a) $-3x^2$; (b) $2x$.

3. (a) $2x^2$; (b) $-3x$.

4. (a) $-x^2$; (b) $-y^{-4}$; (c) y^{-1}; (d) y^{-7}.

Exercises 6.3

1. (a) $4x + 4$; (b) $-4x - 4$; (c) $4x - 4$;

(d) $-4x + 4$.

2. (a) $5x - 5y$; (b) $19x + 57y$; (c) $8a + 8b$;

(d) $5y + xy$; (e) $12x + 48$;

(f) $17x - 153$; (g) $-a + 2b$; (h) $x + \frac{1}{2}$;

(i) $6m - 12m^2 - 9mn$.

3. (a) $18 - 13x - 26 = -8 - 13x$; (b) $x^2 + xy$ will not simplify any further.

4. (a) $x^2 + 7x + 6$; (b) $x^2 + 9x + 20$;

(c) $x^2 + x - 6$; (d) $x^2 + 5x - 6$;

(e) $xm + ym + nx + yn$;

(f) $12 + 3y + 4x + yx$; (g) $25 - x^2$;

(h) $51x^2 - 79x - 10$.

5. (a) $x^2 - 4x - 21$; (b) $6x^2 + 11x - 7$;

(c) $16x^2 - 1$; (d) $x^2 - 9$; (e) $6 + x - 2x^2$.

6. (a) $\frac{29}{2}x - \frac{5}{2}y$; (b) $\frac{5}{4}x + \frac{5}{4}$.

7. (a) $-x + y$; (b) $-a - 2b$; (c) $-\frac{3}{2}p - \frac{1}{2}q$.

8. $(x + 1)(x + 2) = x^2 + 3x + 2$. So

$(x + 1)(x + 2)(x + 3) = (x^2 + 3x + 2)(x + 3)$

$= (x^2 + 3x + 2)(x) + (x^2 + 3x + 2)(3)$

$= x^3 + 3x^2 + 2x + 3x^2 + 9x + 6$

$= x^3 + 6x^2 + 11x + 6$.

Solutions to Chapter 7

Exercises 7.1

1. (a) $9x + 27$; (b) $-5x + 10$; (c) $\frac{1}{2}x + \frac{1}{2}$;

(d) $-a + 3b$; (e) $1/(2x + 2y)$; (f) $x/(yx - y^2)$.

2. (a) $4x^2$ has factors $1, 2, 4, x, 2x, 4x, x^2, 2x^2,$ $4x^2$; (b) $6x^3$ has factors $1, 2, 3, 6, x, 2x, 3x,$ $6x, x^2, 2x^2, 3x^2, 6x^2, x^3, 2x^3, 3x^3, 6x^3$

3. (a) $10(x+2y)$; (b) $3(4a+b)$; (c) $2x(2-3y)$;
   (d) $7(a+2)$;
   (e) $5(2m-3)$; (f) $1/[5(a+7b)]$;
   (g) $1/[5a(a+7b)]$.
4. (a) $3x(5x+1)$; (b) $x(4x-3)$; (c) $4x(x-2)$;
   (d) $3(5-x^2)$;
   (e) $5x^2(2x+1+3y)$; (f) $6ab(a-2b)$;
   (g) $8b(2ac-ba+3c)$.

Exercises 7.2

1. (a) $(x+2)(x+1)$; (b) $(x+7)(x+6)$;
   (c) $(x+5)(x-3)$;
   (d) $(x+10)(x-1)$; (e) $(x-8)(x-3)$;

(f) $(x-10)(x+10)$;
(g) $(x+2)(x+2)$ or $(x+2)^2$.
2. (a) $(2x+1)(x-3)$; (b) $(3x+1)(x-2)$;
   (c) $(5x+3)(2x+1)$;
   (d) $2x^2+12x+16$ has a common factor of 2 which should be written outside a bracket to give $2(x^2+6x+8)$. The bracket can then be factorized to give $2(x+4)(x+2)$.
3. (a) $(x-2)(x+5)$; (b) $(2x+5)(x-4)$;
   (c) $(3x+1)(3x-1)$; (d) $10x^2+14x-12$ has a common factor of 2 which is written outside a bracket to give $2(5x^2+7x-6)$, the bracket can then be factorized to give $2(5x-3)(x+2)$; (e) $(x+13)(x+2)$;
   (f) $(-x+1)(x+3)$.

Solutions to Chapter 8

Exercises 8.2

1. (a) $\dfrac{1}{3+2x}$ (b) $1+2x$ (c) $\dfrac{1}{2+7x}$ (d) $\dfrac{x}{2+7x}$
   (e) $\dfrac{x}{1+7x}$

2. (a) $\dfrac{1}{3+2x}$ (b) $\dfrac{2}{x+7}$ (c) $\dfrac{2(x+4)}{(x-2)(x+4)} = \dfrac{2}{x-2}$
   (d) $\dfrac{7}{ab+9}$ (e) $\dfrac{y}{y+1}$

3. (a) $\dfrac{1}{x-2}$ (b) $\dfrac{2(x-2)}{(x-2)(x+3)} = \dfrac{2}{x+3}$
   (c) $\dfrac{1}{x+2}$ (d) $\dfrac{(x+1)(x+1)}{(x+1)(x-3)} = \dfrac{x+1}{x-3}$

Exercises 8.3

1. (a) $\dfrac{a}{20}$ (b) $\dfrac{5a}{4b}$ (c) $\dfrac{1}{2ab}$ (d) $\dfrac{6x^2}{y^3}$ (e) $\dfrac{3b}{5a^2}$ (f) $\dfrac{x}{4y}$
   (g) $\dfrac{x}{3(x+y)}$ (h) $\dfrac{1}{3(x+4)}$

2. (a) $\dfrac{3y}{z^3}$ (b) $\dfrac{3+x}{y}$ (c) $\dfrac{x}{12}$ (d) $\dfrac{b}{c}$

3. (a) $\dfrac{1}{x+4}$ (b) $\dfrac{4(x-2)}{x}$ (c) $\dfrac{12ab}{5ef} \times \dfrac{f}{4ab^2} = \dfrac{3}{5eb}$
   (d) $\dfrac{x+3y}{2x} \times \dfrac{4x^2}{y} = \dfrac{2x(x+3y)}{y}$ (e) $\dfrac{9}{xyz}$

Exercises 8.4

1. (a) $\dfrac{3}{x}+\dfrac{4}{y} = \dfrac{3y}{xy}+\dfrac{4x}{xy} = \dfrac{3y+4x}{xy}$

   (b) $\dfrac{3}{x^2}+\dfrac{4y}{x} = \dfrac{3}{x^2}+\dfrac{4xy}{x^2} = \dfrac{3+4xy}{x^2}$

   (c) $\dfrac{4ab}{x}+\dfrac{3ab}{2y} = \dfrac{8aby}{2xy}+\dfrac{3abx}{2xy} = \dfrac{8aby+3abx}{2xy}$

   (d) $\dfrac{4xy}{a}+\dfrac{3xy}{2b} = \dfrac{8xyb}{2ab}+\dfrac{3xya}{2ab} = \dfrac{8xyb+3xya}{2ab}$

   (e) $\dfrac{3}{x}-\dfrac{6}{2x} = \dfrac{3}{x}-\dfrac{3}{x} = 0$

   (f) $\dfrac{3x}{2y}-\dfrac{7y}{4x} = \dfrac{6x^2}{4xy}-\dfrac{7y^2}{4xy} = \dfrac{6x^2-7y^2}{4xy}$

   (g) $\dfrac{3}{x+y}-\dfrac{2}{y} = \dfrac{3y}{(x+y)y}-\dfrac{2(x+y)}{(x+y)y}$
   $= \dfrac{3y-2x-2y}{(x+y)y} = \dfrac{y-2x}{(x+y)y}$

   (h) $\dfrac{1}{a+b}-\dfrac{1}{a-b}$
   $= \dfrac{a-b}{(a+b)(a-b)}-\dfrac{a+b}{(a+b)(a-b)}$
   $= \dfrac{a-b-a-b}{(a+b)(a-b)} = \dfrac{-2b}{(a+b)(a-b)}$

   (i) $2x+\dfrac{1}{2x} = \dfrac{4x^2}{2x}+\dfrac{1}{2x} = \dfrac{4x^2+1}{2x}$

(j) $2x - \dfrac{1}{2x} = \dfrac{4x^2}{2x} - \dfrac{1}{2x} = \dfrac{4x^2 - 1}{2x}$

2. (a) $\dfrac{x}{y} + \dfrac{3x^2}{z} = \dfrac{xz}{yz} + \dfrac{3x^2 y}{yz} = \dfrac{xz + 3x^2 y}{yz}$

(b) $\dfrac{4}{a} + \dfrac{5}{b} = \dfrac{4b + 5a}{ab}$

(c) $\dfrac{6x}{y} - \dfrac{2y}{x} = \dfrac{6x^2 - 2y^2}{xy}$

(d) $3x - \dfrac{3x + 1}{4} = \dfrac{12x}{4} - \dfrac{3x + 1}{4}$

$= \dfrac{12x - 3x - 1}{4} = \dfrac{9x - 1}{4}$

(e) $\dfrac{5a}{12} + \dfrac{9a}{18} = \dfrac{15a + 18a}{36} = \dfrac{33a}{36} = \dfrac{11a}{12}$

(f) $\dfrac{x - 3}{4} + \dfrac{3}{5} = \dfrac{5(x - 3) + 12}{20}$

$= \dfrac{5x - 15 + 12}{20} = \dfrac{5x - 3}{20}$

Solutions to Chapter 9

Exercises 9.1

1. (a) $x = \dfrac{y}{3}$ (b) $x = \dfrac{1}{y}$ (c) $x = \dfrac{5 + y}{7}$

(d) $x = 2y + 14$ (e) $x = \dfrac{1}{2y}$

(f) $2x + 1 = \dfrac{1}{y}$ so $2x = \dfrac{1}{y} - 1 = \dfrac{1 - y}{y}$

finally, $x = \dfrac{1 - y}{2y}$

(g) $y - 1 = \dfrac{1}{2x}$ so $2x = \dfrac{1}{y - 1}$

and then $x = \dfrac{1}{2(y - 1)}$

(h) $x = \dfrac{y + 21}{18}$ (i) $x = \dfrac{19 - y}{8}$

2. (a) $m = \dfrac{y - c}{x}$ (b) $x = \dfrac{y - c}{m}$ (c) $c = y - mx$

3. (a) $y = 13x - 26$ and so $x = \dfrac{y + 26}{13}$

(b) $y = x + 1$ and so $x = y - 1$

(c) $y = a + tx - 3t$ and so $tx = y + 3t - a$

and then $x = \dfrac{y + 3t - a}{t}$

4. (a) $x = \dfrac{y - 11}{7}$ (b) $I = \dfrac{V}{R}$ (c) $r^3 = \dfrac{3V}{4\pi}$

and so $r = \sqrt[3]{\dfrac{3V}{4\pi}}$ (d) $m = \dfrac{F}{a}$

5. $n = (l - a)/d + 1$.

6. $\sqrt{x} = (m - n)/t$ and so $x = [(m - n)/t]^2$.

7. (a) $x^2 = 1 - y$ and so $x = \pm\sqrt{1 - y}$;

(b) $1 - x^2 = 1/y$ and so $x^2 = 1 - 1/y$,

finally, $x = \pm\sqrt{1 - 1/y}$; (c) Write

$(1 + x^2)y = 1 - x^2$, remove brackets to get

$y + x^2 y = 1 - x^2$, rearrange and factorize to

give $x^2(y + 1) = 1 - y$ from which

$x^2 = (1 - y)/(1 + y)$, finally

$x = \pm\sqrt{(1 - y)/(1 + y)}$.

Solutions to Chapter 10

Exercises 10.1

1. (a) $x = \dfrac{9}{3} = 3$; (b) $x = 3 \times 9 = 27$;

(c) $3t = -6$, so $t = \dfrac{-6}{3} = -2$; (d) $x = 22$;

(e) $3x = 4$ and so $x = \dfrac{4}{3}$; (f) $x = 36$;

(g) $5x = 3$ and so $x = \dfrac{3}{5}$; (h) $x + 3 = 6$

and so $x = 3$;

(i) adding the two terms on the left we obtain

$$\frac{3x+2}{2} + 3x = \frac{3x+2}{2} + \frac{6x}{2}$$

$$= \frac{3x+2+6x}{2} = \frac{9x+2}{2}$$

and the equation becomes $(9x+2)/2 = 1$, therefore $9x+2 = 2$, that is, $9x = 0$ and so $x = 0$.

2. (a) $5x + 10 = 13$, so $5x = 3$ and so $x = \frac{3}{5}$;
   (b) $3x - 21 = 2x + 2$ and so $x = 23$;
   (c) $5 - 10x = 8 - 4x$ so that $-3 = 6x$ finally $x = \frac{-3}{6} = \frac{1}{2}$.
3. (a) $t = 9$; (b) $v = \frac{17}{7}$; (c) $3s + 2 = 14s - 14$, so that $11s = 16$ and so $s = \frac{16}{11}$.

Exercises 10.2

1. (a) Adding the two equations gives

$$\begin{array}{rcl} 3x + y & = & 1 \\ 2x - y & - & 2 \\ \hline 5x & = & 3 \end{array}$$

from which $x = \frac{3}{5}$. Substitution into either equation gives $y = -\frac{4}{5}$.
(b) Subtracting the given equations eliminates y to give $x = 4$. From either equation $y = 1$.
(c) Multiplying the second equation by 2 and subtracting the first gives

$$\begin{array}{rcl} 2x - y & = & 17 \\ 2x + 6y & = & 24 \\ \hline 7y & = & 7 \end{array} -$$

from which $y = 1$. Substitute into either equation to get $x = 9$.
(d) Multiplying the second equation by -2 and subtracting from the first gives

$$\begin{array}{rcl} -2x + y & = & -21 \\ -2x - 6y & = & 28 \\ \hline 7y & = & -49 \end{array} -$$

from which $y = -7$. Substitution into either equation gives $x = 7$.
(e) Multiplying the first equation by 3 and adding the second gives

$$\begin{array}{rcl} -3x + 3y & = & -30 \\ 3x + 7y & = & 20 \\ \hline 10y & = & -10 \end{array} +$$

from which $y = -1$. Substitution into either equation gives $x = 9$.
(f) Multiplying the second equation by 2 and subtracting this from the first gives

$$\begin{array}{rcl} 4x - 2y & = & 2 \\ 6x - 2y & = & 8 \\ \hline -2x & = & -6 \end{array} -$$

from which $x = 3$. Substitution into either equation gives $y = 5$.

Exercises 10.3

1. (a) $(x+2)(x-1) = 0$ so that $x = -2$ and 1;
   (b) $(x-3)(x-5) = 0$ so that $x = 3$ and 5;
   (c) $2(2x+1)(x+1) = 0$ so that $x = -1$ and $x = -\frac{1}{2}$; (d) $(x-3)(x-3) = 0$ so that $x = 3$ twice;
   (e) $(x+9)(x-9) = 0$ so that $x = -9$ and 9.

2. (a) $x = \dfrac{-(-6) \pm \sqrt{(-6)^2 - 4(3)(-5)}}{6}$

   $= \dfrac{6 \pm \sqrt{96}}{6}$

   $= 2.633$ and -0.633

   (b) $x = \dfrac{-3 \pm \sqrt{9 - 4(1)(-77)}}{2}$

   $= \dfrac{-3 \pm \sqrt{317}}{2} = 7.402$ and -10.402

   (c) $x = \dfrac{-(-9) \pm \sqrt{(-9)^2 - 4(2)(2)}}{4}$

   $= \dfrac{9 \pm \sqrt{65}}{4} = 4.266$ and 0.234

3. (a) $(3x+2)(2x+3) = 0$
   from which $x = -2/3$ and $-3/2$;

(b) $(t+3)(3t+4) = 0$
from which $t = -3$ and $-4/3$;

(c) $t = (7 \pm \sqrt{37})/2 = 6.541$ and 0.459.

Solutions to Chapter 11

Exercises 11.2

1. (a) Multiply the input by 10; (b) multiply the input by -1 and then add 2, alternatively we could say subtract the input from 2;
   (c) raise the input to the power 4 and then multiply the result by 3; (d) divide 4 by the square of the input; (e) take three times the square of the input, subtract twice the input from the result and finally add 9; (f) the output is always 5 whatever the value of the input; (g) the output is always zero whatever the value of the input.

2. (a) Take three times the square of the input and add to twice the input; (b) take three times the square of the input and add to twice the input. The functions in parts (a) and (b) are the same. Both instruct us to do the same thing.

3. (a) $f(x) = x^3/12$ – letters other than f and x are also valid;
   (b) $f(x) = (x+3)^2$; (c) $f(x) = x^2 + 4x - 10$;
   (d) $f(x) = x/(x^2 + 5)$; (e) $f(x) = x^3 - 1$;
   (f) $f(x) = (x-1)^2$; (g) $f(x) = (7 - 2x)/4$;
   (h) $f(x) = -13$.

4. (a) $A(2) = 3$; (b) $A(3) = 7$; (c) $A(0) = 1$;
   (d) $A(-1) = (-1)^2 - (-1) + 1 = 3$.

5. (a) $y(1) = 1$;
   (b) $y(-1) = [2(-1) - 1]^2 = (-3)^2 = 9$;
   (c) $y(-3) = 49$; (d) $y(0.5) = 0$;
   (e) $y(-0.5) = 4$.

6. (a) $f(t+1) = 4(t+1) + 6 = 4t + 10$;
   (b) $f(t+2) = 4(t+2) + 6 = 4t + 14$;
   (c) $f(t+1) - f(t) = (4t+10) - (4t+6) = 4$;
   (d) $f(t+2) - f(t) = (4t+14) - (4t+6) = 8$.

7. (a) $f(n) = 2n^2 - 3$; (b) $f(z) = 2z^2 - 3$;
   (c) $f(t) = 2t^2 - 3$;
   (d) $f(2t) = 2(2t)^2 - 3 = 8t^2 - 3$;
   (e) $f(1/z) = 2(1/z)^2 - 3 = 2/z^2 - 3$;
   (f) $f(3/n) = 2(3/n)^2 - 3 = 18/n^2 - 3$;
   (g) $f(-x) = 2(-x)^2 - 3 = 2x^2 - 3$;
   (h) $f(-4x) = 2(-4x)^2 - 3 = 32x^2 - 3$;
   (i) $f(x+1) = 2(x+1)^2 - 3 = 2(x^2 + 2x + 1) - 3 = 2x^2 + 4x - 1$;
   (j) $f(2x-1) = 2(2x-1)^2 - 3 = 8x^2 - 8x - 1$.

8. $a(p+1) = (p+1)^2 + 3(p+1) + 1 = p^2 + 2p + 1 + 3p + 3 + 1 = p^2 + 5p + 5$
   $a(p+1) - a(p) = (p^2 + 5p + 5) - (p^2 + 3p + 1) = 2p + 4$.

9. (a) $f(3) = 6$; (b) $h(2) = 3$;
   (c) $f[h(2)] = f(3) = 6$; (d) $h[f(3)] = h(6) = 7$.

10. (a) $f[h(t)] = f(t+1) = 2(t+1) = 2t + 2$;
    (b) $h[f(t)] = h(2t) = 2t + 1$. Note that $h[f(t)]$ is not equal to $f[h(t)]$.

11. (a) $f(0.5) = 0.5$ (first part); (b) $f(1.1) = 1$ (third part); (c) $f(1) = 2$ (second part).

Exercises 11.3

1. In each case let the inverse be denoted by g
   (a) $g(x) = x/3$;
   (b) $g(x) = 4x$; (c) $g(x) = x - 1$;
   (d) $g(x) = x + 3$; (e) $g(x) = 3 - x$;
   (f) $g(x) = (x-6)/2$; (g) $g(x) = (7-x)/3$;
   (h) $g(x) = 1/x$;
   (i) $g(x) = 3/x$; (j) $g(x) = -3/4x$.

Solutions to Chapter 12

Exercises 12.1
1.

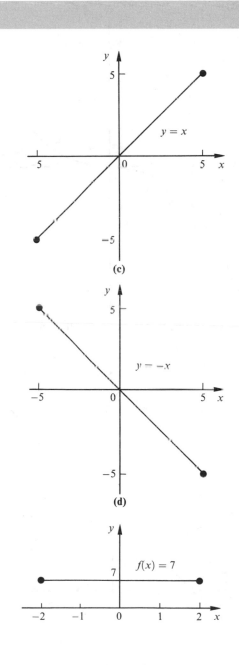

2. All y coordinates must be the same.
3. All x coordinates must be the same.

Exercises 12.2
1.

2.

3. (a)

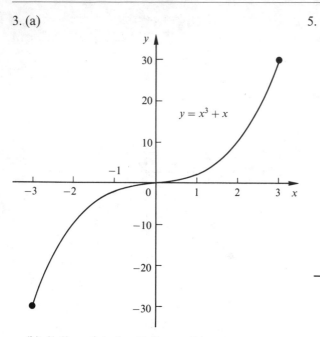

$$y = x^3 + x$$

(b) $(1,2)$ and $(-1,-2)$ lie on the curve.

4.

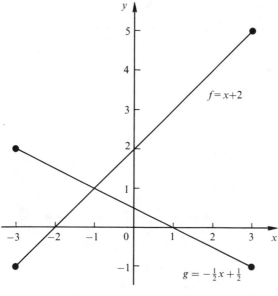

$f = x + 2$

$g = -\frac{1}{2}x + \frac{1}{2}$

Graphs intersect at $(-1,1)$

5.

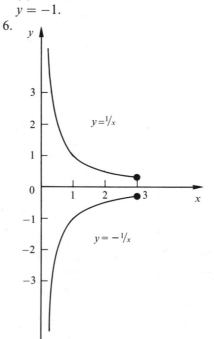

$y = x^2 - x - 1$

(a) The curve cuts the horizontal axis at $x = -0.6$ and $x = 1.6$;
(b) the curve cuts the vertical axis at $y = -1$.

6.

$y = 1/x$

$y = -1/x$

7.

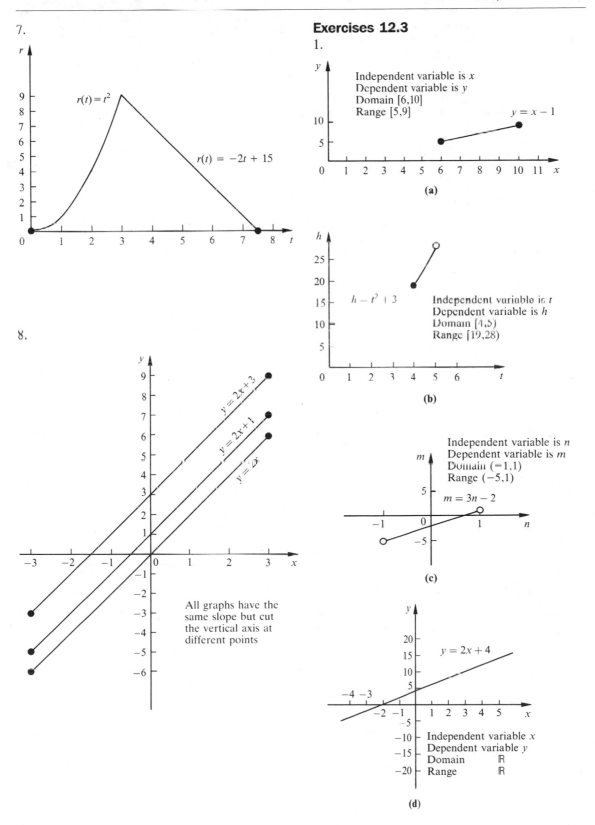

Exercises 12.3

1.

Independent variable is x
Dependent variable is y
Domain [6,10]
Range [5,9]

$y = x - 1$

$r(t) = t^2$

$r(t) = -2t + 15$

(a)

$h - t^2 + 3$

Independent variable is t
Dependent variable is h
Domain [4,5)
Range [19,28)

(b)

8.

Independent variable is n
Dependent variable is m
Domain (−1,1)
Range (−5,1)

$m = 3n - 2$

(c)

$y = 2x+3$

$y = 2x+1$

$y = 2x$

All graphs have the
same slope but cut
the vertical axis at
different points

$y = 2x + 4$

Independent variable x
Dependent variable y
Domain $\mathbb{R}$
Range $\mathbb{R}$

(d)

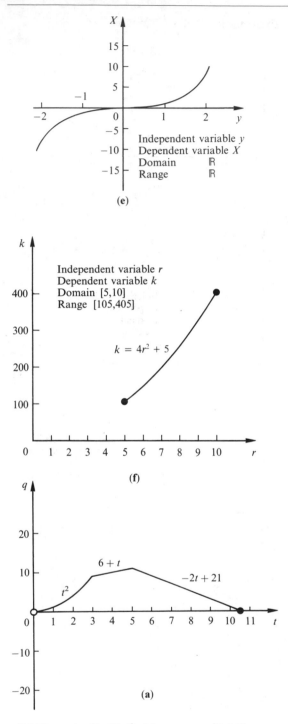

(e)

(f)

(a)

(b) Domain $(0, 10.5]$; (c) range $= [0, 11]$;
(d) $q(1) = 1$, $q(3) = 9$, $q(5) = 11$, $q(7) = 7$.

Exercises 12.4

1.

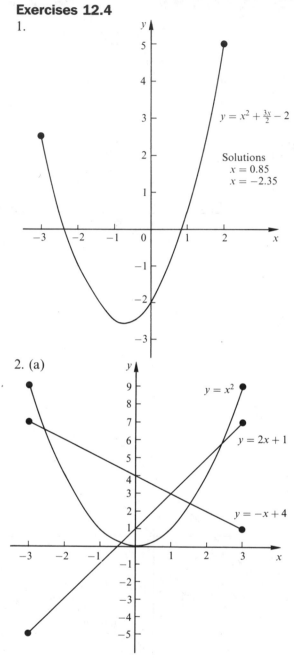

$y = x^2 + \frac{3x}{2} - 2$

Solutions
$x = 0.85$
$x = -2.35$

2. (a)

(b) Solutions of $x^2 = 2x + 1$ are $x = -0.4$
and $x = 2.4$;
(c) $x^2 + x - 4 = 0$ can be written as
$x^2 = -x + 4$, solutions are $x = 1.6$ and
$x = -2.6$.

Solutions to Chapter 13

Exercises 13.1
1. (a), (b), (c) and (e) are straight lines.
2. (a) $m = 9$, $c = -11$, (b) $m = 8$, $c = 1.4$,
   (c) $m = \frac{1}{2}$, $c = -11$;
   (d) $m = -2$, $c = 17$, (e) $m = \frac{2}{3}$, $c = \frac{1}{3}$,
   (f) $m = -\frac{2}{5}$, $c = \frac{4}{5}$,
   (g) $m = 3$, $c = -3$, (h) $m = 0$, $c = 4$.

Exercises 13.2
1. $y = 2x + 5$.

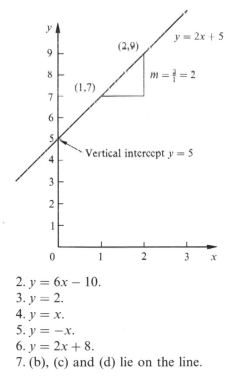

2. $y = 6x - 10$.
3. $y = 2$.
4. $y = x$.
5. $y = -x$.
6. $y = 2x + 8$.
7. (b), (c) and (d) lie on the line.

Exercises 13.3
1.

2.

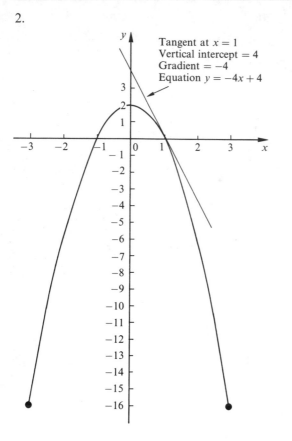

Tangent at $x = 1$
Vertical intercept $= 4$
Gradient $= -4$
Equation $y = -4x + 4$

Solutions to Chapter 14

Exercises 14.1

1. (a) 9.9742; (b) 6.6859; (c) 0.5488; (d) 0.8808;
   (e) -8.4901.

2. (a) $e^2 . e^7 = e^{2+7} = e^9$; (b) $e^7/e^4 = e^{7-4} = e^3$;
   (c) $e^{-2}/e^3 = e^{-2-3} = e^{-5}$ or $1/e^5$;
   (d) $e^3/e^{-1} = e^{3-(-1)} = e^4$;
   (e) $(4e)^2/(2e)^3 = 4^2 e^2/2^3 e^3 = 16 e^{2-3}/8$
   $= 2 e^{-1} = 2/e$;
   (f) $e^2(e + 1/e) - e = e^2 e + e^2/e - e$
   $= e^3 + e - e = e^3$;
   (g) $e^{1.5} e^{2.7} = e^{1.5+2.7} = e^{4.2}$;
   (h) $\sqrt{e}\sqrt{4e} = e^{0.5}(4e)^{0.5} = e^{0.5} 4^{0.5} e^{0.5}$
   $= 2 e^1 = 2e$.

3. (a) $e^t/e^3 = e^{t-3}$;
   (b) $2e^{3x}/4e^x = \frac{1}{2}e^{3x-x} = e^{2x}/2$;
   (c) $(e^x)^3/3e^x = e^{3x}/3e^x = e^{3x-x}/3 = e^{2x}/3$;
   (d) $e^x + 2e^{-x} + e^x = 2e^x + 2e^{-x}$
   $= 2(e^x + e^{-x})$;
   (e) $e^x e^y e^z/e^{(x/2)-y} = e^{x+y+z-((x/2)-y)}$
   $= e^{(x/2)+2y+z}$;
   (f) $(e^{t/3})^6/(e^t + e^t)$
   $= e^{(t/3)\times 6}/2e^t = e^{2t}/2e^t = e^t/2$;
   (g) $e^t/e^{-t} = e^{t-(-t)} = e^{2t}$;
   (h) $(1 + e^{-t})/e^{-t} = (1 + e^{-t})e^t = e^t + e^{-t}e^t$
   $= e^t + e^0 = e^t + 1$;

(i) $(e^z e^{-z/2})^2 = (e^{z-z/2})^2 = (e^{z/2})^2 = e^z$;

(j) $e^{-3+t}/2e^t = e^{-3+t-t}/2 = e^{-3}/2 = 1/2e^3$;

(k) $(e^{-1}/e^{-x})^{-1} = (e^{-1-(-x)})^{-1} = (e^{x-1})^{-1}$
$= e^{(x-1)(-1)} = e^{1-x}$

4. (a) $\sinh x + \cosh x = \dfrac{e^x - e^{-x}}{2} + \dfrac{e^x + e^{-x}}{2}$

$= \dfrac{e^x - e^{-x} + e^x + e^{-x}}{2}$

$= \dfrac{2e^x}{2} = e^x$

(b) $\cosh x - \sinh x = \dfrac{e^x + e^{-x}}{2} - \left(\dfrac{e^x - e^{-x}}{2}\right)$

$= \dfrac{e^x + e^{-x} - e^x + e^{-x}}{2}$

$= \dfrac{2e^{-x}}{2} = e^{-x}$

(c) $(\cosh x)^2 - (\sinh x)^2 =$
$= (\cosh x + \sinh x)(\cosh x - \sinh x)$
$= e^x . e^{-x} = 1$

Exercises 14.2

1.

x	-1	-0.8	-0.6	-0.4	-0.2
e^{3x}	0.0498	0.0907	0.1653	0.3012	0.5488

x	0	0.2	0.4	0.6	0.8	1
e^{3x}	1	1.8221	3.3201	6.0496	11.0232	20.0855

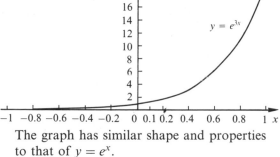

The graph has similar shape and properties to that of $y = e^x$.

2. (a)

t	0	1	2	3	4	5
P	5	8.1606	9.3233	9.7511	9.9084	9.9663

$P = 10 - 5e^{-t}$

(b) Maximum population is 10, obtained for large values of t.

Exercises 14.3

1. (a)

x	0	0.5	1	1.5	2	2.5	3
$15 - x^2$	15	14.75	14	12.75	11	8.75	6
e^x	1	1.65	2.72	4.48	7.39	12.18	20.09

$y = e^x$

$y = 15 - x^2$

(b) Solving $e^x + x^2 = 15$ is equivalent to solving $e^x = 15 - x^2$. Approximate solution is $x = 2.3$.

2. (a)

x	−1	−0.75	−0.5	−0.25
$12x^2 - 1$	11	5.75	2	−0.25
$3e^{-x}$	8.15	6.35	4.95	3.85

x	0	0.25	0.5	0.75	1
$12x^2 - 1$	−1	−0.25	2	5.75	11
$3e^{-x}$	3	2.34	1.82	1.42	1.10

(b) Approximate solutions are $x = -0.80$ and $x = 0.49$.

3.

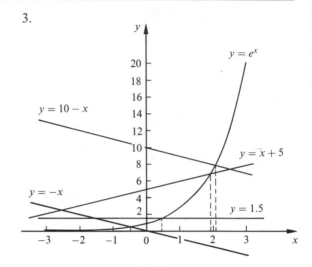

(a) The equation $e^x + x = 0$ is equivalent to $e^x = -x$. Hence draw $y = -x$. Intersection of $y = e^x$ and $y = -x$ will give solution to $e^x + x = 0$. Approximate solution: $x = -0.57$.

(b) The equation $e^x - 1.5 = 0$ is equivalent to $e^x = 1.5$. Hence also draw $y = 1.5$. This is a horizontal straight line. Intersection point gives solution to $e^x - 1.5 = 0$. Approximate solution: $x = 0.41$.

(c) The equation $e^x - x - 5 = 0$ can be written as $e^x = x + 5$. Hence draw $y = x + 5$. Approximate solution: $x = 1.94$.

(d) The equation

$$\frac{e^x}{2} + \frac{x}{2} - 5 = 0$$

can be written as

$$e^x = 10 - x$$

Hence also draw $y = 10 - x$. Approximate solution: $x = 2.07$.

Solutions to Chapter 15

Exercises 15.1

1. (a) 2.1761; (b) 5.0106; (c) −0.5003;
   (d) −2.3026.

2. (a) $\log_3 6561 = 8$; (b) $\log_6 7776 = 5$;
   (c) $\log_2 1024 = 10$;
   (d) $\log_{10} 100000 = 5$; (e) $\log_4 16384 = 7$;
   (f) $\log_{0.5} 0.03125 = 5$;
   (g) $\log_{12} 1728 = 3$; (h) $\log_9 6561 = 4$.

3. (a) $1296 = 6^4$; (b) $225 = 15^2$; (c) $512 = 8^3$;
   (d) $2401 = 7^4$;
   (e) $243 = 3^5$; (f) $216 = 6^3$; (g) $8000 = 20^3$;
   (h) $4096 = 16^3$;
   (i) $4096 = 2^{12}$.

Exercises 15.2

1. (a) $\log_4 6 = \log 6 / \log 4 = 1.2925$;
   (b) $\log_3 10 = \log 10 / \log 3 = 2.0959$;
   (c) $\log_{20} 270 = \log 270 / \log 20 = 1.8688$;
   (d) $\log_5 0.65 = \log 0.65 / \log 5 = -0.2677$;
   (e) $\log_2 100 = \log 100 / \log 2 = 6.6439$;
   (f) $\log_2 0.03 = \log 0.03 / \log 2 = -5.0589$;
   (g) $\log_{100} 10 = \log 10 / \log 100 = \frac{1}{2}$;
   (h) $\log_7 7 = 1$.

Exercises 15.3

1. (a) $\log 5 + \log 9 = \log(5 \times 9) = \log 45$;
   (b) $\log 9 - \log 5 = \log(\frac{9}{5}) = \log 1.8$;
   (c) $\log 5 - \log 9 = \log(\frac{5}{9})$;
   (d) $2\log 5 + \log 1 = 2\log 5 = \log 5^2 = \log 25$;
   (e) $2\log 4 - 3\log 2 = \log 4^2 - \log 2^3$
   $= \log 16 - \log 8 = \log(\frac{16}{8}) = \log 2$;
   (f) $\log 64 - 2\log 2 = \log(64/2^2) = \log 16$;
   (g) $3\log 4 + 2\log 1 + \log 27 - 3\log 12$
   $$= \log 4^3 + 2(0) + \log 27 - \log 12^3$$
   $$= \log\left(\frac{4^3 \times 27}{12^3}\right)$$
   $$= \log 1 = 0$$

2. (a) $\log 4x - \log x = \log(4x/x) = \log 4$;
   (b) $\log t^3 + \log t^4 = \log(t^3 \times t^4) = \log t^7$;

(c) $\log 2t - \log\left(\dfrac{t}{4}\right) = \log\left(2t \div \dfrac{t}{4}\right)$

$$= \log\left(2t \times \dfrac{4}{t}\right) = \log 8$$

(d) $\log 2 + \log\left(\dfrac{3}{x}\right) - \log\left(\dfrac{x}{2}\right)$

$$= \log\left(2 \times \dfrac{3}{x}\right) - \log\left(\dfrac{x}{2}\right)$$

$$= \log\left(\dfrac{6}{x} \div \dfrac{x}{2}\right)$$

$$= \log\left(\dfrac{6}{x} \times \dfrac{2}{x}\right)$$

$$= \log\left(\dfrac{12}{x^2}\right)$$

(e) $\log\left(\dfrac{t^2}{3}\right) + \log\left(\dfrac{6}{t}\right) - \log\left(\dfrac{1}{t}\right)$

$$= \log\left(\dfrac{t^2}{3} \times \dfrac{6}{t}\right) - \log\left(\dfrac{1}{t}\right)$$

$$= \log(2t) - \log\left(\dfrac{1}{t}\right)$$

$$= \log\left(\dfrac{2t}{1/t}\right)$$

$$= \log 2t^2$$

(f) $2\log y - \log y^2 = \log y^2 - \log y^2 = 0$;

(g) $3\log\left(\dfrac{1}{t}\right) + \log t^2 = \log\left(\dfrac{1}{t}\right)^3 + \log t^2$

$$= \log\left(\dfrac{1}{t^3}\right) + \log t^2$$

$$= \log\left(\dfrac{t^2}{t^3}\right)$$

$$= \log\left(\dfrac{1}{t}\right) = -\log t$$

(h) $4\log\sqrt{x} + 2\log\left(\dfrac{1}{x}\right) = 4\log x^{0.5} + \log\left(\dfrac{1}{x}\right)^2$

$\qquad\qquad = \log x^2 + \log\left(\dfrac{1}{x^2}\right)$

$\qquad\qquad = \log\left(x^2\,\dfrac{1}{x^2}\right)$

$\qquad\qquad = \log 1 = 0$

(i) $2\log x + 3\log t = \log x^2 + \log t^3 = \log(x^2 t^3)$;

(j) $\log A - \dfrac{1}{2}\log 4A = \log A - \log(4A)^{1/2}$

$\qquad\qquad = \log\left[\dfrac{A}{(4A)^{1/2}}\right]$

$\qquad\qquad = \log\left(\dfrac{A^{1/2}}{2}\right)$

(k) $\dfrac{\log 9x + \log 3x^2}{3} = \dfrac{\log(9x.3x^2)}{3}$

$\qquad\qquad = \dfrac{\log(27x^3)}{3}$

$\qquad\qquad = \dfrac{\log(3x)^3}{3}$

$\qquad\qquad = \dfrac{3\log(3x)}{3}$

$\qquad\qquad = \log(3x)$

(l) $\log xy + 2\log\left(\dfrac{x}{y}\right) + 3\log\left(\dfrac{y}{x}\right)$

$\qquad = \log xy + \log\left(\dfrac{x^2}{y^2}\right) + \log\left(\dfrac{y^3}{x^3}\right)$

$\qquad = \log\left(xy\,\dfrac{x^2}{y^2}\,\dfrac{y^3}{x^3}\right)$

$\qquad = \log\left(\dfrac{x^3 y^4}{x^3 y^2}\right)$

$\qquad = \log y^2$

(m) $\log\left(\dfrac{A}{B}\right) - \log\left(\dfrac{B}{A}\right)$

$= \log\left(\dfrac{A}{B}\div\dfrac{B}{A}\right) = \log\left(\dfrac{A}{B}\times\dfrac{A}{B}\right) = \log\left(\dfrac{A^2}{B^2}\right)$

(n) $\log\left(\dfrac{2t}{3}\right) + \dfrac{1}{2}\log 9t - \log\left(\dfrac{1}{t}\right)$

$\qquad = \log\left(\dfrac{2t}{3}\right) + \log(9t)^{1/2} - \log t^{-1}$

$\qquad = \log\left[\dfrac{2t}{3}(9t)^{1/2}\right] + \log t$

$\qquad = \log\left[\dfrac{2t}{3}(9t)^{1/2}t\right]$

$\qquad = \log\left(\dfrac{2t}{3}3t^{1/2}t\right)$

$\qquad = \log(2t^{2.5})$

Exercises 15.4

1. (a) $\log x = 0.3940$

   $\qquad x = 10^{0.3940} = 2.4774$

   (b) $\ln x = 0.3940$

   $\qquad x = e^{0.3940} = 1.4829$

   (c) $10^y = 5.5$

   $\qquad y = \log 5.5 = 0.7404$

   (d) $e^z = 500$

   $\qquad z = \ln 500 = 6.2146$

   (e) $\log(3v) = 1.6512$

   $\qquad 3v = 10^{1.6512}$

   $\qquad v = \dfrac{10^{1.6512}}{3} = 14.9307$

   (f) $\ln\left(\dfrac{t}{6}\right) = 1$

   $\qquad \dfrac{t}{6} = e^1 = e$

   $\qquad t = 6e = 16.3097$

   (g) $10^{2r+1} = 25$

   $\qquad 2r + 1 = \log 25$

   $\qquad 2r = \log 25 - 1$

   $\qquad r = \dfrac{\log 25 - 1}{2} = 0.1990$

(h) $e^{(2t-1)/3} = 7.6700$

$$\frac{2t-1}{3} = \ln 7.6700$$

$$2t - 1 = 3\ln 7.6700$$

$$2t = 3\ln 7.6700 + 1$$

$$t = \frac{3\ln 7.6700 + 1}{2} = 3.5560$$

(i) $\log(4b^2) = 2.6987$

$$4b^2 = 10^{2.6987}$$

$$b^2 = \frac{10^{2.6987}}{4} = 124.9223$$

$$b = \sqrt{124.9223} = \pm 11.1769$$

(j) $\log\left(\frac{6}{2+t}\right) = 1.5$

$$\frac{6}{2+t} = 10^{1.5}$$

$$6 = 10^{1.5}(2+t)$$

$$\frac{6}{10^{1.5}} = 2 + t$$

$$t = \frac{6}{10^{1.5}} - 2 = -1.8103$$

(k) $\ln(2r^3 + 1) = 3.0572$

$$2r^3 + 1 = e^{3.0572}$$

$$2r^3 = e^{3.0572} - 1$$

$$r^3 = \frac{e^{3.0572} - 1}{2} = 10.1340$$

$$r = 2.1640$$

(l) $\ln(\log t) = -0.3$

$$\log t = e^{-0.3} = 0.7408$$

$$t = 10^{0.7408} = 5.5058$$

(m) $10^{t^2-1} = 180$

$$t^2 - 1 = \log 180$$

$$t^2 = \log 180 + 1 = 3.2553$$

$$t = \sqrt{3.2553} = \pm 1.8042$$

(n) $10^{3r^4} = 170000$

$$3r^4 = \log 170000$$

$$r^4 = \frac{\log 170000}{3} = 1.7435$$

$$r = \pm 1.1491$$

(o) $\log 10^t = 1.6$

$$t \log 10 = 1.6$$

$$t = 1.6$$

(p) $\ln(e^x) = 20000$

$$x \ln e = 20000$$

$$x = 20000$$

2. (a) $e^{3x}e^{2x} = 59$

$$e^{5x} = 59$$

$$5x = \ln 59$$

$$x = \frac{\ln 59}{5} = 0.8155$$

(b) $10^{3t}.10^{4-t} = 27$

$$10^{4+2t} = 27$$

$$4 + 2t = \log 27$$

$$2t = \log 27 - 4$$

$$t = \frac{\log 27 - 4}{2} = -1.2843$$

(c) $\log(5-t) + \log(5+t) = 1.2$

$$\log(5-t)(5+t) = 1.2$$

$$\log(25 - t^2) = 1.2$$

$$25 - t^2 = 10^{1.2}$$

$$t^2 = 25 - 10^{1.2}$$

$$= 9.1511$$

$$t = \pm 3.0251$$

(d) $\log x + \ln x = 4$

$$\log x + \frac{\log x}{\log e} = 4$$

$$\log x \left(1 + \frac{1}{\log e}\right) = 4$$

$$3.3026 \log x = 4$$

$$\log x = \frac{4}{3.3026} = 1.2112$$

$$x = 10^{1.2112} = 16.2619$$

(b)

x	0.1	0.2	0.5	0.75	1
$\log(1/x)$	1	0.6990	0.3010	0.1249	0

x	2	3	4	5	6
$\log(1/x)$	−0.3010	−0.4771	−0.6021	−0.6990	−0.7782

x	7	8	9	10
$\log(1/x)$	−0.8451	−0.9031	−0.9542	−1

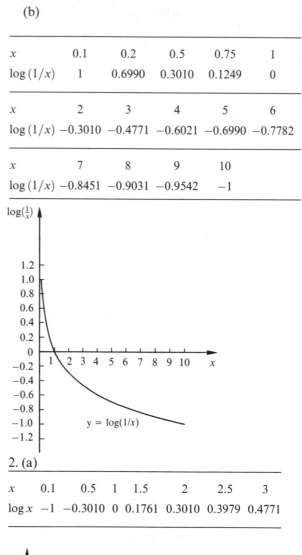

$y = \log(1/x)$

Exercises 15.5

1. (a)

x	0.5	1	2	3	4
$\log x^2$	−0.6021	0	0.6021	0.9542	1.2041

x	5	6	7	8	9	10
$\log x^2$	1.3979	1.5563	1.6902	1.8062	1.9085	2

$y = \log x^2$

2. (a)

x	0.1	0.5	1	1.5	2	2.5	3
$\log x$	−1	−0.3010	0	0.1761	0.3010	0.3979	0.4771

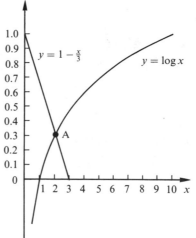

$y = 1 - \frac{x}{3}$ $y = \log x$

A

(b) $y = 1 - x/3$ is a straight line passing through $(0,1)$ and $(3,0)$.

(c) The graphs intersect at A. Approximate solution: $x = 2.06$.

Solutions to Chapter 16

Exercises 16.1

1. (a) $240° = \pi/180 \times 240 = 4.19^c$;
   (b) $300° = \pi/180 \times 300 = 5.24^c$;
   (c) $400° = \pi/180 \times 400 = 6.98^c$;
   (d) $37° = \pi/180 \times 37 = 0.65^c$;
   (e) $1000° = \pi/180 \times 1000 = 17.45^c$.
2. (a) $\pi/10^c = 180/10 = 18°$;
   (b) $2\pi/9^c = \frac{2}{9} \times 180 = 40°$;
   (c) $2^c = 180/\pi \times 2 = 114.59°$;
   (d) $3.46^c = 180/\pi \times 3.46 = 198.24°$;
   (e) $1.75^c = 180/\pi \times 1.75 = 100.27°$.

Exercises 16.2

1. (a) length of arc $= r\theta = 15(2) = 30$ cm;
   (b) length of arc $= r\theta = 15(3) = 45$ cm;
   (c) length of arc $= r\theta = 15(1.2) = 18$ cm;
   (d) length of arc $= r\theta = 15(100 \times \pi/180)$
       $= 26.18$ cm;
   (e) length of arc $= r\theta = 15(217 \times \pi/180)$
       $= 56.81$ cm.
2. (a) angle subtended
   $= \theta = \dfrac{\text{arc length}}{\text{radius}} = \dfrac{12}{12} = 1^c$

   (b) $\theta = \dfrac{\text{arc length}}{\text{radius}} = \dfrac{6}{12} = 0.5^c$

   (c) $\theta = \dfrac{\text{arc length}}{\text{radius}} = \dfrac{24}{12} = 2^c$

   (d) $\theta = \dfrac{\text{arc length}}{\text{radius}} = \dfrac{15}{12} = 1.25^c$

(e) $\theta = \dfrac{\text{arc length}}{\text{radius}} = \dfrac{2}{12} = \left(\dfrac{1}{6}\right)^c$

Exercises 16.3

1. (a)
   area of sector $= \dfrac{r^2\theta}{2} = \dfrac{9^2(1.5)}{2} = 60.75$ cm^2

   (b) area of sector $= \dfrac{r^2\theta}{2} = \dfrac{9^2(2)}{2} = 81$ cm^2

   (c) area of sector $= \dfrac{r^2\theta}{2} = \dfrac{9^2}{2}\left(100 \times \dfrac{\pi}{180}\right)$
       $= 70.69$ cm^2

   (d) area of sector $= \dfrac{r^2\theta}{2} = \dfrac{9^2}{2}\left(215 \times \dfrac{\pi}{180}\right)$
       $= 151.97$ cm^2

2. area of sector $= \dfrac{r^2\theta}{2}$

   $\theta = \dfrac{2 \times \text{area}}{r^2}$

   When $r = 16$ we have

   $\theta = \dfrac{\text{area}}{128}$

   (a) Angle $= \theta = \frac{100}{128} = 0.78^c$;
   (b) angle $= \frac{5}{128} = 0.039^c$;
   (c) angle $= \frac{520}{128} = 4.06^c$.

Solutions to Chapter 17

Exercises 17.1

1. (a) 0.5774; (b) 0.3420; (c) 0.2588.

2. (a) $\sin 70° = 0.9397$, $\cos 20° = 0.9397$;
   (b) $\sin 25° = 0.4226$, $\cos 65° = 0.4226$;

(c) in Figure 17.1, $A = \theta$ and $C = 90 - \theta$, hence

$\sin\theta = BC/AC$ and $\cos(90 - \theta) = BC/AC$, so $\sin\theta = \cos(90 - \theta)$.

3. (a) $\sin\theta = BC/AC = 3/\sqrt{18}$;

(b) $\cos\theta = AB/AC = 3/\sqrt{18}$;

(c) $\tan\theta = BC/AB = \frac{3}{3} = 1$.

Exercises 17.2

1. When $\sin\alpha > 0$ then α must be in the first or second quadrant. When $\cos\alpha < 0$ then α must be in the second or third quadrant. To satisfy both conditions, α must be in the second quadrant.

2. β can be in the third or fourth quadrant.

3. (a) $\theta = 50°$. $\theta = 180 - 50 = 130°$;

(b) $\theta = 180 - 40 = 140°$ and also $\theta = 180 + 40 = 220°$;

(c) $\theta = 20°$. $\theta = 180 + 20 = 200°$;

(d) $\theta = 180 + 70 = 250°$, $\theta = 360 - 70 = 290°$;

(e) $\theta = 10°$. $\theta = 360 - 10 = 350°$;

(f) $\theta = 180 - 80 = 100°$, $\theta = 360 - 80 = 280°$;

(g) $\theta = 0°$. $\theta = 180°$, $360°$;

(h) $\theta = 90°$. $\theta = 360 - 90 = 270°$.

Exercises 17.3

1. (a)

θ	0	30	60	90	120	150
$\sin(\theta + 45°)$	0.71	0.97	0.97	0.71	0.26	−0.26

θ	180	210	240	270	300	330	360
$\sin(\theta + 45°)$	−0.71	−0.97	−0.97	−0.71	−0.26	0.26	0.71

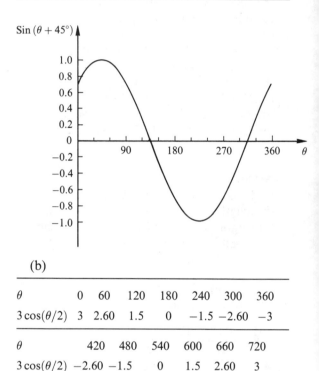

(b)

θ	0	60	120	180	240	300	360
$3\cos(\theta/2)$	3	2.60	1.5	0	−1.5	−2.60	−3

θ	420	480	540	600	660	720
$3\cos(\theta/2)$	−2.60	−1.5	0	1.5	2.60	3

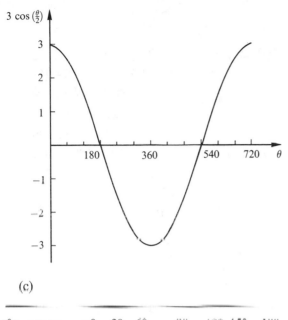

3 cos $\left(\frac{\theta}{2}\right)$

(c)

θ	0	30	60	90	120	150	180
$2\tan(\theta + 60°)$	3.46	–	−3.46	−1.15	0	1.15	3.46

θ		210	240	270	300	330	360
$2\tan(\theta + 60°)$	–		−3.46	−1.15	0	1.15	3.46

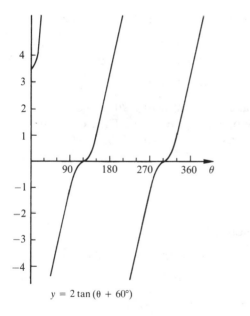

$y = 2\tan(\theta + 60°)$

Exercises 17.4

1. (a) Using a calculator, we see
$\sin^{-1}(0.9) = 64.16°$.
Also $\sin^{-1}(0.9) = 180 - 64.16 = 115.84°$.
(b) Using a calculator we see
$\cos^{-1}(0.45) = 63.26°$.
Also $\cos^{-1}(0.45) = 360 - 63.26 = 296.74°$.
(c) Using a calculator we see
$\tan^{-1} 1.3 = 52.43°$.
Also $\tan^{-1} 1.3 = 180 + 52.43 = 232.43°$.
(d) Using a calculator we see
$\sin^{-1} 0.6 = 36.87$. Hence
$\sin^{-1}(-0.6) = 180 + 36.87 = 216.87°$ and
$360 - 36.87 = 323.13°$.
(e) Using a calculator we see
$\cos^{-1} 0.75 = 41.41°$. So
$\cos^{-1}(-0.75) = 180 - 41.41 = 138.59°$ and
$180 + 41.41 = 221.41°$.
(f) We have $\tan^{-1} 0.3 = 16.70°$ and so
$\tan^{-1}(-0.3) = 180 - 16.70 = 163.30°$ and
$360 - 16.70 = 343.30°$.
(g) $\sin^{-1} 1 = 90°$.
(h) $\cos^{-1}(-1) = 180°$.

Exercises 17.5

1. (a) $\sin(2\theta + 50°) = 0.5$
$$2\theta + 50° = (30), 150, 390, 510, 750, ...$$
$$2\theta = (-20), 100, 340, 460, 700, ...$$
$$\theta = 50, 170, 230, 350, ...$$

Solutions in the range 0° to 360° are
$$\theta = 50°, \ 170°, \ 230°, \ 350°$$

(b) $\cos(\theta + 110°) = 0.3$
$$\theta + 110° = (72.54), 287.46, 432.54$$
$$\theta = 177.46°, \ 322.54°$$

(c) $\tan\left(\dfrac{\theta}{2}\right) = 1$

$\qquad \dfrac{\theta}{2} = 45°, (225°)$

$\qquad \theta = 90°$

(d) $\sin\left(\dfrac{\theta}{2}\right) = -0.5$

$\qquad \dfrac{\theta}{2} = 210, 330$

$\qquad \theta = 420, 660$

No solution between $0°$ and $360°$.

(e) $\cos(2\theta - 30°) = -0.5$

$\qquad 2\theta - 30° = 120, 240, 480, 600$

$\qquad 2\theta = 150, 270, 510, 630$

$\qquad \theta = 75°, \ 135°, \ 255°, \ 315°$

(f) $\tan(3\theta - 20°) = 0.25$

$\qquad 3\theta - 20° = 14.04, 194.04, 374.04,$

$\qquad\qquad 554.04, 734.04, 914.04$

$\qquad 3\theta = 34.04, 214.04, 394.04,$

$\qquad\qquad 574.04, 754.04, 934.04$

$\qquad \theta = 11.35°, \ 71.35°, \ 131.35°,$

$\qquad\qquad 191.35°, \ 251.35°, \ 311.35°$

(g) $\sin 2\theta = 2\cos 2\theta$

$\qquad \dfrac{\sin 2\theta}{\cos 2\theta} = 2$

$\qquad \tan 2\theta = 2$

$\qquad 2\theta = 63.43, 243.43, 423.43, 603.43$

$\qquad \theta = 31.72°, \ 121.72°, \ 211.72°, \ 301.72°$

Solutions to Chapter 18

Exercises 18.1

1. $C = 180 - 70 - 42 = 68°.$

2. $A + B + C = 180$

$\qquad B + C = 180 - A = 180 - 42 = 138$

$\qquad 2C + C = 138$

$\qquad\qquad C = 46°$

$\qquad\qquad B = 2C = 92°$

3. $A + B + C = 180$

$\qquad A + B = 180 - C = 180 - 114 = 66$

Since A and B are equal then $A = B = 33°$.

4. In $\triangle ABC$, $A = 50°$, $B = 72°$ and $C = 58°$:
(a) the largest angle is B and so the longest side is b, that is, AC; (b) the smallest angle is A and so the shortest side is a, that is, BC.

5. There are two solutions.

(i) Suppose $A = 40°$ and $B = C$. Then

$\qquad A + B + C = 180°$

$\qquad\qquad B + C = 180 - 40 = 140$

$\qquad\qquad\quad B = C = 70°$

The angles are $40°, \ 70°, \ 70°$.
(ii) Suppose $A = B = 40°$. Then $C = 100°$.
The angles are $40°, \ 40°, \ 100°$.

Exercises 18.2

1. $a^2 = c^2 - b^2 = 30^2 - 17^2 = 611$

$\qquad a = \sqrt{611} = 24.72$

The length of BC is 24.72 cm.

2. $d^2 = c^2 + e^2$

$$= (1.7)^2 + (1.2)^2 = 4.33$$

$$d = \sqrt{4.33} - 2.08$$

The length of CE is 2.08 cm.

3. Let the length of $b = AC$ be k cm. Then $AC/AB = b/c = k/c = \frac{2}{1}$ so $c = k/2$. The length of $c = AB$ is $k/2$ cm. Now, $a = BC$ and $b^2 = a^2 + c^2$

$$k^2 = a^2 + \left(\frac{k}{2}\right)^2 = a^2 + \frac{k^2}{4}$$

$$\frac{3}{4}k^2 = a^2$$

$$a = \sqrt{\frac{3}{4}k^2} - \frac{\sqrt{3}}{2}k$$

So $AC:BC = k : (\sqrt{3}/2)k$. This is the same as $1 : \sqrt{3}/2$ or $2 : \sqrt{3}$.

4. Let the length of $y = XZ$ be k cm. The length of YZ is then $2k$ cm. Using Pythagoras's theorem we see that

$$x^2 = y^2 + z^2$$

$$(2k)^2 = k^2 + 10^2$$

$$4k^2 = k^2 + 100$$

$$3k^2 = 100$$

$$k^2 = \frac{100}{3}$$

$$k = \sqrt{\frac{100}{3}} = 5.774$$

$$2k = 11.547$$

The length of XZ is 5.774 cm and the length of YZ is 11.547 cm.

5. $m^2 = l^2 + n^2$

$$30^2 = l^2 + 26^2$$

$$l^2 = 30^2 - 26^2 = 224$$

$$l = \sqrt{224} = 14.97$$

The length of MN is 14.97 cm.

Exercises 18.3

1. $B = 180 - 90 - 37 = 53°$.

$$\tan A = \frac{BC}{AC} = \frac{a}{36}$$

$$\tan 37° = \frac{a}{36}$$

$$a = 36 \tan 37° = 27.13$$

$$\cos A = \frac{AC}{AB} = \frac{36}{c}$$

$$c = \frac{36}{\cos A} = \frac{36}{\cos 37°} = 45.08$$

The solution is

$A = 37°$ $a = BC = 27.13$ cm
$B = 53°$ $b = AC = 36$ cm
$C = 90°$ $c = AB = 45.08$ cm

2. Using Pythagoras's theorem

$$e^2 = c^2 + d^2$$

$$= 21^2 + 14^2 = 637$$

$$e = \sqrt{637} = 25.24$$

$$\tan C = \frac{21}{14} = 1.5$$

$$C = \tan^{-1}(1.5) = 56.31°$$

Finally, $D = 180 - 90 - 56.31 = 33.69°$. The solution is

$C = 56.31°$ $c = DE = 21$ cm
$D = 33.69°$ $d = CE = 14$ cm
$E = 90°$ $e = CD = 25.24$ cm

3. $Z = 180 - 90 - 26 = 64°$

$$\sin Y = \sin 26° = \frac{XZ}{YZ} = \frac{y}{45}$$

$$y = 45 \sin 26° = 19.73$$

and

$$\cos Y = \cos 26° = \frac{XY}{YZ} = \frac{z}{45}$$

$$z = 45 \cos 26° = 40.45$$

The solution is

$X = 90°$ $x = YZ = 45$ mm
$Y = 26°$ $y = XZ = 19.73$ mm
$Z = 64°$ $z = XY = 40.45$ mm

4. Using Pythagoras's theorem,

$$j^2 = i^2 + k^2$$
$$= 27^2 + 15^2 = 954$$
$$j = \sqrt{954} = 30.89$$

$$\tan I = \frac{i}{k} = \frac{27}{15} = 1.8$$
$$I = \tan^{-1} 1.8 = 60.95°$$

and

$$\tan K = \frac{15}{27}$$
$$K = \tan^{-1}\left(\frac{15}{27}\right) = 29.05°$$

The solution is
$$I = 60.95° \quad i = JK = 27 \text{ cm}$$
$$J = 90° \quad j = IK = 30.89 \text{ cm}$$
$$K = 29.05° \quad k = IJ = 15 \text{ cm}$$

5. $Q = 180 - 62 - 28 = 90°$ and so $\triangle PQR$ is a right-angled triangle.

$$\sin R = \frac{r}{q} = \frac{r}{22}$$
$$PQ = r = 22 \sin R = 22 \sin 28° = 10.33$$

and

$$\cos R = \frac{p}{q} = \frac{p}{22}$$
$$QR = p = 22 \cos R = 22 \cos 28° = 19.42$$

The solution is
$$P = 62° \quad p = QR = 19.42 \text{ cm}$$
$$Q = 90° \quad q = PR = 22 \text{ cm}$$
$$R = 28° \quad r = PQ = 10.33 \text{ cm}$$

6. $T = 180 - 70 - 20 = 90°$ so $\triangle RST$ is a right-angled triangle.

$$\cos R = \cos 70° = \frac{s}{t} = \frac{12}{t}$$
$$RS = t = \frac{12}{\cos 70°} = 35.09$$

$$\tan R = \tan 70° = \frac{r}{s} = \frac{r}{12}$$
$$ST = r = 12 \tan 70° = 32.97$$

The solution is
$$R = 70° \quad r = ST = 32.97 \text{ cm}$$
$$S = 20° \quad s = RT = 12 \text{ cm}$$
$$T = 90° \quad t = RS = 35.09 \text{ cm}$$

Exercises 18.4

1. (a) $\dfrac{a}{\sin A} = \dfrac{b}{\sin B} = \dfrac{c}{\sin C}$

$$\frac{a}{\sin A} = \frac{24}{\sin B} = \frac{31}{\sin 37°}$$

So

$$\sin B = \frac{24 \sin 37°}{31} = 0.4659$$
$$B = \sin^{-1}(0.4659) = 27.77°$$

Hence $A = 180 - 37 - 27.77 = 115.23°$. Finally

$$a = \frac{31 \sin A}{\sin 37°} = \frac{31 \sin 115.23°}{\sin 37°} = 46.60$$

The solution is
$$A = 115.23° \quad a = BC = 46.60 \text{ cm}$$
$$B = 27.77° \quad b = AC = 24 \text{ cm}$$
$$C = 37° \quad c = AB = 31 \text{ cm}$$

(b) Substituting the given values into the sine rule gives

$$\frac{a}{\sin A} = \frac{24}{\sin B} = \frac{19}{\sin 37°}$$

So

$$\sin B = \frac{24 \sin 37°}{19} = 0.7602$$
$$B = \sin^{-1}(0.7602) = 49.48° \text{ or } 130.52°$$

Both values are acceptable and so there are 2 solutions.

Case 1: $B = 49.48°$

$$A = 180 - 37 - 49.48 = 93.52°$$
$$a = \frac{19 \sin A}{\sin 37°} = \frac{19 \sin 93.52}{\sin 37°} = 31.51$$

Case 2: $B = 130.52°$

$A = 180 - 37 - 130.52 = 12.48°$

$a = \dfrac{19 \sin A}{\sin 37°} = \dfrac{19 \sin 12.48°}{\sin 37°} = 6.82$

The solutions are
$A = 93.52°$ $a = BC = 31.51$ cm
$B = 49.48°$ $b = AC = 24$ cm
$C = 37°$ $c = AB = 19$ cm
and
$A = 12.48°$ $a = BC = 6.82$ cm
$B = 130.52°$ $b = AC = 24$ cm
$C = 37°$ $c = AB = 19$ cm

(c) $\dfrac{17}{\sin 17°} = \dfrac{b}{\sin B} = \dfrac{c}{\sin 101°}$

$c = \dfrac{17 \sin 101°}{\sin 17°} = 57.08$

Also $B = 180 - 17 - 101 = 62°$. So

$b = \dfrac{17 \sin B}{\sin 17°} = \dfrac{17 \sin 62°}{\sin 17°} = 51.34$

The solution is
$A = 17°$ $a = BC = 17$ cm
$B = 62°$ $b = AC = 51.34$ cm
$C = 101°$ $c = AB = 57.08$ cm

(d) $\dfrac{8.9}{\sin 53°} = \dfrac{b}{\sin B} = \dfrac{9.6}{\sin C}$

$\sin C = \dfrac{9.6 \sin 53°}{8.9} = 0.8614$

$C = \sin^{-1}(0.8614) = 59.48°$ or $120.52°$

Both values are acceptable and so there are 2 solutions.
Case 1: $C = 59.48°$

$B = 180 - 53 - 59.48 = 67.52°$

$b = \dfrac{8.9 \sin B}{\sin 53°} = \dfrac{8.9 \sin 67.52°}{\sin 53°} = 10.30$

Case 2: $C = 120.52°$

$B = 180 - 53 - 120.52 = 6.48°$

$b = \dfrac{8.9 \sin B}{\sin 53°} = \dfrac{8.9 \sin 6.48°}{\sin 53°} = 1.26$

The solutions are
$A = 53°$ $a = BC = 8.9$ cm
$B = 67.52°$ $b = AC = 10.30$ cm

$C = 59.48°$ $c = AB = 9.6$ cm
and
$A = 53°$ $a = BC = 8.9$ cm
$B = 6.48°$ $b = AC = 1.26$ cm
$C = 120.52°$ $c = AB = 9.6$ cm

(e) $\dfrac{14.5}{\sin 62°} = \dfrac{b}{\sin B} = \dfrac{12.2}{\sin C}$

$\sin C = \dfrac{12.2 \sin 62°}{14.5} = 0.7429$

$C = \sin^{-1}(0.7429) = 47.98°$

Hence $B = 180 - 62 - 47.98 = 70.02°$.

$b = \dfrac{14.5 \sin B}{\sin 62°} = \dfrac{14.5 \sin 70.02°}{\sin 62°} = 15.43$

The solution is
$A = 62°$ $a = BC = 14.5$ cm
$B = 70.02°$ $b = AC = 15.43$ cm
$C = 47.98°$ $c = AB = 12.2$ cm

(f) $\dfrac{a}{\sin A} = \dfrac{11}{\sin 36°} = \dfrac{c}{\sin 50°}$

$c = \dfrac{11 \sin 50°}{\sin 36°} = 14.34$

Also $A = 180 - 36 - 50 = 94°$ so that

$a = \dfrac{11 \sin A}{\sin 36°} = \dfrac{11 \sin 94°}{\sin 36°} = 18.67$

The solution is
$A = 94°$ $a = BC = 18.67$ cm
$B = 36°$ $b = AC = 11$ cm
$C = 50°$ $c = AB = 14.34$ cm

Exercises 18.5

1. (a) $a = BC = 86$, $b = AC = 119$, $c = AB = 53$.

$$a^2 = b^2 + c^2 - 2bc \cos A$$
$$86^2 = 119^2 + 53^2 - 2(119)(53) \cos A$$
$$12614 \cos A = 119^2 + 53^2 - 86^2 = 9574$$
$$\cos A = \dfrac{9574}{12614} = 0.7590$$
$$A = \cos^{-1}(0.7590) = 40.62°$$

$$b^2 = a^2 + c^2 - 2ac\cos B$$
$$119^2 = 86^2 + 53^2 - 2(86)(53)\cos B$$
$$9116\cos B = 86^2 + 53^2 - 119^2 = -3956$$
$$\cos B = -\frac{3956}{9116} = -0.4340$$
$$B = \cos^{-1}(-0.4340) = 115.72°$$

$C = 180 - 40.62 - 115.72 = 23.66°$. The solution is

$A = 40.62°$ $a = BC = 86$ cm
$B = 115.72°$ $b = AC = 119$ cm
$C = 23.66°$ $c = AB = 53$ cm

(b) $b = AC = 30$, $c = AB = 42$, $A = 115°$.

$$a^2 = b^2 + c^2 - 2bc\cos A$$
$$a^2 = 30^2 + 42^2 - 2(30)(42)\cos 115°$$
$$= 3728.9980$$
$$a = \sqrt{3728.998} = 61.07$$

Now using the cosine rule again,
$$b^2 = a^2 + c^2 - 2ac\cos B$$
$$30^2 = 3728.998 + 1764$$
$$- 2(61.07)(42)\cos B$$
$$5129.88\cos B = 4592.998$$
$$\cos B = 0.8953$$
$$B = 26.45°$$

$C = 180 - 115 - 26.45 = 38.55°$. The solution is

$A = 115°$ $a = BC = 61.07$ cm
$B = 26.45°$ $b = AC = 30$ cm
$C = 38.55°$ $c = AB = 42$ cm

(c) $a = BC = 74$, $b = AC = 93$, $C = 39°$.

$$c^2 = a^2 + b^2 - 2ab\cos C$$
$$c^2 = 74^2 + 93^2 - 2(74)(93)\cos 39°$$
$$= 3428.3630$$
$$c = 58.552$$

$$a^2 = b^2 + c^2 - 2bc\cos A$$
$$5476 = 8649 + 3428.3630$$
$$- 2(93)(58.552)\cos A$$
$$10890.672\cos A = 6601.363$$
$$\cos A = 0.6061$$
$$A = 52.69°$$

$B = 180 - 52.69 - 39 = 88.31°$. The solution is

$A = 52.69°$ $a = BC = 74$ cm
$B = 88.31°$ $b = AC = 93$ cm
$C = 39°$ $c = AB = 58.55$ cm

(d) $a = BC = 3.6$, $b = AC = 2.7$, $c = AB = 1.9$.
$$a^2 = b^2 + c^2 - 2bc\cos A$$
$$3.6^2 = 2.7^2 + 1.9^2 - 2(2.7)(1.9)\cos A$$
$$10.26\cos A = -2.06$$
$$\cos A = -0.2008$$
$$A = 101.58°$$

$$b^2 = a^2 + c^2 - 2ac\cos B$$
$$2.7^2 = 3.6^2 + 1.9^2 - 2(3.6)(1.9)\cos B$$
$$13.68\cos B = 3.6^2 + 1.9^2 - 2.7^2 = 9.28$$
$$\cos B = \frac{9.28}{13.68} = 0.6784$$
$$B = 47.28°$$

$C = 180 - 101.58 - 47.28 = 31.14°$. The solution is

$A = 101.58°$ $a = BC = 3.6$ cm
$B = 47.28°$ $b = AC = 2.7$ cm
$C = 31.14°$ $c = AB = 1.9$ cm

(e) $a = BC = 39$, $c = AB = 29$, $B = 100°$.
$$b^2 = a^2 + c^2 - 2ac\cos B$$
$$= 39^2 + 29^2 - 2(39)(29)\cos 100°$$
$$= 2754.7921$$
$$b = 52.49$$

$$a^2 = b^2 + c^2 - 2bc \cos A$$
$$39^2 = 52.49^2 + 29^2 - 2(52.49)(29) \cos A$$
$$3044.42 \cos A = 52.49^2 + 29^2 - 39^2 = 2074.79$$
$$\cos A = \frac{2074.79}{3044.42} = 0.6815$$
$$A = 47.04°$$

$C = 180 - 100 - 47.04 = 32.96°$. The solution is

$$A = 47.04° \quad a = BC = 39 \text{ cm}$$
$$B = 100° \quad b = AC = 52.49 \text{ cm}$$
$$C = 32.96° \quad c = AB = 29 \text{ cm}$$

Solutions to Chapter 19

Exercises 19.1

1. A is 1×4; B is 4×2; C is 1×1, and D is 5×1.

2. A is square; B is neither diagonal nor square; C is square and diagonal (it is the 3×3 identity matrix); D and E are neither square nor diagonal; F is square and diagonal; G and H are square.

3. A 3×1 matrix has 3 elements, a 1×3 matrix has 3 elements. An $m \times n$ matrix has mn elements and an $n \times n$ matrix has n^2 elements.

Exercises 19.2

1. $MN = \begin{pmatrix} 13 & 46 \\ 7 & 0 \end{pmatrix} \quad NM = \begin{pmatrix} 25 & 22 \\ 1 & -12 \end{pmatrix}$

Note $MN \neq NM$.

2. (a) $3A = \begin{pmatrix} 3 & -21 \\ -3 & 6 \end{pmatrix}$

(b) $A + B$ not possible; (c) $A + C$ not possible;

(d) $5D = \begin{pmatrix} 35 \\ 45 \\ 40 \\ 5 \end{pmatrix}$

(c) BC not possible; (f) CB not possible;

(g) AC not possible; (h) CA not possible;

(i) $5A - 2C$ not possible;

(j) $D - E = \begin{pmatrix} 6.5 \\ 9 \\ 8 \\ 0 \end{pmatrix}$

(k) BA not possible;

(l) $AB = \begin{pmatrix} 10 & 7 & 72 \\ -5 & -7 & -22 \end{pmatrix}$

(m) $A^2 = AA = \begin{pmatrix} 8 & -21 \\ -3 & 11 \end{pmatrix}$

3. (a) $IA = \begin{pmatrix} 3 & 7 \\ -1 & -2 \end{pmatrix} = A$

(b) $AI = A$; (c) BI not possible;

(d) $IB = \begin{pmatrix} x \\ y \end{pmatrix} = B$

(e) CI not possible; (f) $IC = C$. Where it is possible to multiply by the identity matrix, the result is unaltered.

4. (a) $A + B = \begin{pmatrix} a+e & b+f \\ c+g & d+h \end{pmatrix}$

(b) $A - B = \begin{pmatrix} a-e & b-f \\ c-g & d-h \end{pmatrix}$

(c) $AB = \begin{pmatrix} ae+bg & af+bh \\ ce+dg & cf+dh \end{pmatrix}$

(d) $BA = \begin{pmatrix} ea+fc & eb+fd \\ ga+hc & gb+hd \end{pmatrix}$

5. $\begin{pmatrix} -7x+5y \\ -2x-y \end{pmatrix}$

6. $\begin{pmatrix} 9a-5b \\ 3a+8b \end{pmatrix}$

Exercises 19.3

1. (a) -2; (b) -22; (c) 1; (d) 88.
2. C and E are singular.

4. (a) $\dfrac{1}{-4}\begin{pmatrix} 10 & -4 \\ -6 & 2 \end{pmatrix} = \begin{pmatrix} \frac{-5}{2} & 1 \\ \frac{3}{2} & \frac{-1}{2} \end{pmatrix}$

(b) $\begin{pmatrix} 1 & 0 \\ 0 & 1 \end{pmatrix}$

(c) $\dfrac{1}{-8}\begin{pmatrix} 2 & -3 \\ -2 & -1 \end{pmatrix} = \begin{pmatrix} \frac{-1}{4} & \frac{3}{8} \\ \frac{1}{4} & \frac{1}{8} \end{pmatrix}$

(d) has no inverse.

Exercises 19.4

1. (a) Take $A = \begin{pmatrix} 1 & 1 \\ 2 & -1 \end{pmatrix}$

so that $A^{-1} = \begin{pmatrix} \frac{1}{3} & \frac{1}{3} \\ \frac{2}{3} & \frac{-1}{3} \end{pmatrix}$

Then $\begin{pmatrix} x \\ y \end{pmatrix} = \begin{pmatrix} \frac{1}{3} & \frac{1}{3} \\ \frac{2}{3} & \frac{-1}{3} \end{pmatrix}\begin{pmatrix} 17 \\ 10 \end{pmatrix} = \begin{pmatrix} 9 \\ 8 \end{pmatrix}$

so that $x = 9$ and $y = 8$.

(b) Take $A = \begin{pmatrix} 2 & -1 \\ 1 & 3 \end{pmatrix}$

so that $A^{-1} = \begin{pmatrix} \frac{3}{7} & \frac{1}{7} \\ \frac{-1}{7} & \frac{2}{7} \end{pmatrix}$

Then $\begin{pmatrix} x \\ y \end{pmatrix} = \begin{pmatrix} \frac{3}{7} & \frac{1}{7} \\ \frac{-1}{7} & \frac{2}{7} \end{pmatrix}\begin{pmatrix} 10 \\ -2 \end{pmatrix} = \begin{pmatrix} 4 \\ -2 \end{pmatrix}$

so that $x = 4$ and $y = -2$.

(c) Take $A = \begin{pmatrix} 1 & 2 \\ 3 & -1 \end{pmatrix}$

so that $A^{-1} = \begin{pmatrix} \frac{1}{7} & \frac{2}{7} \\ \frac{3}{7} & \frac{-1}{7} \end{pmatrix}$

Then $\begin{pmatrix} x \\ y \end{pmatrix} = \begin{pmatrix} \frac{1}{7} & \frac{2}{7} \\ \frac{3}{7} & \frac{-1}{7} \end{pmatrix}\begin{pmatrix} 15 \\ 10 \end{pmatrix} = \begin{pmatrix} 5 \\ 5 \end{pmatrix}$

so that $x = y = 5$.

2. The matrix

$$\begin{pmatrix} 2 & -3 \\ 6 & -9 \end{pmatrix}$$

is singular and so has no inverse. Hence the matrix method fails to produce any solutions. The second equation is simply a multiple of the first – that is, we really only have one equation and this is not sufficient information to produce a unique value of x and y.

Solutions to Chapter 20

Exercises 20.1
1. 16 cm $=$ 160 mm $=$ 0.16 m.
2. 356 mm $=$ 35.6 cm $=$ 0.356 m.
3. 156 m $=$ 0.156 km.
4. $A = 14 \times 7 = 98$ cm^2.
5. $A = 17.5$ m^2, cost $=$ £87.33.
6. $A = 2 \times 8 = 16$ m^2.
7. $s = (a + b + c)/2 = 17$,
   $A = \sqrt{s(s-a)(s-b)(s-c)}$
   $= \sqrt{2295} = 47.9$ cm^2.
8. $A = \frac{1}{2}(a+b)h = 22$ cm^2.
9. $A = \pi r^2 = 1017.9$ cm^2.
10. $A = 78.5$ cm^2.
11. $A - 26 = \pi r^2$ and so $r^2 - 26/\pi = 8.276$,
    then $r = 2.877$ cm.
12. 13430 cm^2 $= 1.3430$ m^2.
13. $V = \pi r^2 h = 1099.6$ cm^3 $- 1.0996$ litres.
14. $V = \frac{4}{3}\pi r^3 = 2145$ cm^3.

15. $86 = \frac{4}{3}\pi r^3$, therefore
    $r^3 = (3 \times 86)/(4 \times \pi) = 20.53$,
    then $r = 2.74$ cm.
16. $A = \pi r^2 \times \theta/360 = 39.27$ cm^2.
17. $V = \pi r^2 h = 28.274$ m^3, now 1000 litres
    $= 1$ m^3
    and so 28.274 m^3 $= 28274$ litres.

Exercises 20.2
1. (a) 12 kg $=$ 12000 g; (b) 12 kg $-$ 12000000 mg.
2. 168 mg $=$ 0.168 g $=$ 0.000168 kg or
   1.68×10^{-4} kg.
3. 0.005 tonnes $=$ 5 kg.
4. 875 tonnes $-$ 875000 kg $=$ 8.75 $\times$ 10^5 kg.
5. 3500 kg $-$ 3.5 tonnes.

Solutions to Chapter 21

Exercises 21.2
1. (a) $8x^7$; (b) $7x^6$; (c) $-x^{-2} = -1/x^2$;
   (d) $-5x^{-6} = -5/x^6$; (e) $13x^{12}$;
   (f) $5x^4$; (g) $-2x^{-3} = -2/x^3$.
2. (a) $\frac{3}{2}x^{1/2}$; (b) $\frac{5}{2}x^{3/2}$; (c) $-\frac{1}{2}x^{-3/2}$;
   (d) $\frac{1}{2}x^{-1/2} = 1/2\sqrt{x}$;
   (e) $y = \sqrt{x} = x^{1/2}$ and so
   $y' = \frac{1}{2}x^{-1/2} = 1/2\sqrt{x}$; (f) $0.2x^{-0.8}$.
3. $y = 8$, $y' = 0$, the line is horizontal and so
   its gradient is zero.

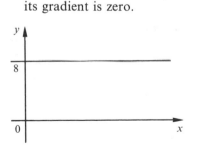

4. (a) $16x^{15}$; (b) $0.5x^{-0.5}$; (c) $-3.5x^{-4.5}$;
   (d) $y = 1/x^{1/2} = x^{-1/2}$ and so $y' = -\frac{1}{2}x^{-3/2}$;
   (e) $y = 1/\sqrt{x} = 1/x^{1/2} = x^{-1/2}$ and
   so $y' = -\frac{1}{2}x^{-3/2}$.
5. $y' = 4x^3$: (a) at $x = -4$, $y' = -256$;
   (b) at $x = -1$, $y' = -4$;
   (c) at $x = 0$, $y' = 0$; (d) at $x = 1$, $y' = 4$;
   (e) at $x = 4$, $y' = 256$. The graph is falling
   at the given negative x values and is rising
   at the given positive x values. At $x = 0$ the
   graph is neither rising nor falling.
6. (a) $y' = 3x^2$, $y'(-3) = 27$, $y'(0) = 0$,
       $y'(3) = 27$;
   (b) $y' = 4x^3$, $y'(-2) = -32$, $y'(2) = 32$;
   (c) $y' = \frac{1}{2}x^{-1/2} = 1/2\sqrt{x}$, $y'(1) = \frac{1}{2}$,
   $y'(2) = \frac{1}{2\sqrt{2}} = 0.354$; (d) $y' = -2x^{-3}$,
   $y'(-2) = \frac{1}{4}$, $y'(2) = -\frac{1}{4}$.
7. (a) $2t$; (b) $-3t^{-4}$; (c) $\frac{1}{2}t^{-1/2}$; (d) $-1.5t^{-2.5}$.

8. (a) $-7\sin 7x$; (b) $\frac{1}{2}\cos(x/2)$; (c) $-\frac{1}{2}\sin(x/2)$;
   (d) $4e^{4x}$; (e) $-3e^{-3x}$.
9. (a) $1/x$; (b) $\frac{1}{3}\cos(x/3)$; (c) $\frac{1}{2}e^{x/2}$.
10. $y' = \cos x$, $y'(0) = 1$, $y'(\pi/2) = 0$,
    $y'(\pi) = -1$.
11. (a) $7t^6$; (b) $3\cos 3t$; (c) $2e^{2t}$; (d) $1/t$;
    (e) $-\frac{1}{3}\sin(t/3)$.

Exercises 21.3

1. (a) $y' = 6x^5 - 4x^3$; (b) $y' = 6 \times (2x) = 12x$;
   (c) $y' = 9(-2x^{-3}) = -18x^{-3}$; (d) $y' = \frac{1}{2}$;
   (e) $y' = 6x^2 - 6x$.
2. (a) $y' = 2\cos x$;
   (b) $y' = 3(4\cos 4x) = 12\cos 4x$;
   (c) $y' = -63\sin 9x + 12\cos 4x$;
   (d) $y' = 3e^{3x} + 10e^{-2x}$.
3. The gradient function is $y' = 9x^2 - 9$.
   $y'(1) = 0$, $y'(0) = -9$, $y'(-1) = 0$. Of the
   three given points, the curve is steepest at
   $x = 0$ where it is falling.
4. When x^2 is differentiated we obtain $2x$.
   Also, because the derivative of any constant
   is zero, the derivative of $x^2 + c$ will be $2x$
   where c can take any value.

Exercises 21.4

1. (a) $y' = 2x$, $y'' = 2$; (b) $y' = 24x^7$,
   $y'' = 168x^6$; (c) $y' = 45x^4$, $y'' = 180x^3$;

(d) $y' = \frac{3}{2}x^{-1/2}$, $y'' = -\frac{3}{4}x^{-3/2}$;
(e) $y' = 12x^3 + 10x$, $y'' = 36x^2 + 10$;
(f) $y' = 27x^2 - 14$, $y'' = 54x$;
(g) $y' = \frac{1}{2}x^{-1/2} - 2x^{-3/2}$,
$y'' = -\frac{1}{4}x^{-3/2} + 3x^{-5/2}$.
2. (a) $y' = 3\cos 3x - 2\sin 2x$,
   $y'' = -9\sin 3x - 4\cos 2x$;
   (b) $y' = e^x - e^{-x}$, $y'' = e^x + e^{-x}$.
3. $y' = 4\cos 4x$, $y'' = -16\sin 4x$, and so
   $16y + y'' = 16\sin 4x + (-16\sin 4x) = 0$ as
   required.
4. $y' = 4x^3 - 6x^2 - 72x$, $y'' = 12x^2 - 12x - 72$.
   The values of x where $y'' = 0$ are found
   from $12x^2 - 12x - 72 = 0$, that is,
   $12(x^2 - x - 6) = 12(x + 2)(x - 3) = 0$ so
   that $x = -2$ and $x = 3$.

Exercises 21.5

1. (a) $y' = 2x$, $y'' = 2$, $(0,1)$ is a minimum;
   (b) $y' = -2x$, $y'' = -2$, $(0,0)$ is a maximum;
   (c) $y' = 6x^2 + 18x$, $y'' = 12x + 18$, $(0,0)$ is a
   minimum and $(-3,27)$ is a maximum;
   (d) $y' = -6x^2 + 54x$, $y'' = -12x + 54$, $(0,0)$
   is a minimum and (9729) is a maximum;
   (e) $y' = 3x^2 - 6x + 3$, $y'' = 6x - 6$,
   stationary point by solving $3x^2 - 6x + 3 = 0$,
   at $(1,2)$, second derivative test fails,
   inspection of gradient on either side of $x = 1$
   reveals a point of inflexion.

Solutions to Chapter 22

Exercises 22.2

1. (a) $x^2/2 + c$; (b) $x^2 + c$; (c) $x^6/6 + c$;
   (d) $x^8/8 + c$; (e) $-x^{-4}/4 + c$;
   (f) $2x^{1/2} + c$; (g) $x + c$; (h) $4x^{3/4}/3 + c$.
2. (a) $9x + c$; (b) $\frac{1}{2}x + c$;
   (c) $-7x + c$; (d) $0.5x + c$. In general,
   $\int k\,dx = kx + c$.

3. (a) $-(\cos 5x)/5 + c$; (b) $-2\cos(x/2) + c$;
   (c) $2\sin(x/2) + c$; (d) $-e^{-3x}/3 + c$;
   (e) $-4\cos 0.25x + c$; (f) $-2e^{-0.5x} + c$;
   (g) $2e^{x/2} + c$.

Exercises 22.3

1. (a) $x^3/3 + x^2/2 + c$

(b) $x^3/3 + x^2/2 + x + c$ (c) $\frac{4}{7}x^7 + c$

(d) $5x^2/2 + 7x + c$ (e) $\ln x + c$

(f) $\int (1/x^2)dx = \int x^{-2}dx = x^{-1}/(-1) + c$

$= -1/x + c$

(g) $3\ln x \quad \dfrac{7}{x} + c$

(h) $\dfrac{2}{3}x^{3/2} + c$

(i) $8x^{1/2} + c$

(j) $\int \dfrac{1}{\sqrt{x}}dx = \int x^{-1/2}dx - 2x^{1/2} + c$

(k) $\dfrac{3x^4}{4} + \dfrac{7x^3}{3} + c$

(l) $\dfrac{2x^5}{5} - x^3 + c$

(m) $\dfrac{x^6}{6} - 7x + c$

2. (a) $3e^x + c$; (b) $\frac{1}{2}e^x + c$; (c) $3e^x + 2e^{-x} + c$;
(d) $5e^{2x}/2 + c$;
(e) $-\cos x - (\sin 3x)/3 + c$.

3. (a) $x^3 - 2e^x + c$; (b) $x^2/4 + x + c$;
(c) $-\frac{2}{3}\cos 3x - (8\sin 3x)/3 + c$;
(d) $-(3\cos 2x)/2 + (5\sin 3x)/3 + c$;
(e) $5\ln x + c$.

4. $5\ln t - t^3/3 + c$.

(d) $\left[-\dfrac{x^3}{3}\right]_{-1}^{1} = -\dfrac{2}{3}$

(e) $\left[\dfrac{x^3}{3}\right]_{0}^{2} = \dfrac{8}{3}$

(f) $\left[\dfrac{x^3}{3} + 2x^2\right]_{0}^{3} = 9 + 18 = 27$

(g) $[\sin x]_0^\pi = 0$

(h) $[\ln x]_1^2 = \ln 2 - \ln 1 = \ln 2 = 0.693$

(i) $[\cos x]_0^\pi = \cos \pi - \cos 0 = -1 - 1 = -2$

(j) $[13x]_1^4 = 52 - 13 = 39$

(k) $\left[\dfrac{x^3}{3} - \cos x\right]_0^\pi$

$-\left(\dfrac{\pi^3}{3} - \cos \pi\right) - (0 - \cos 0)$

$-\dfrac{\pi^3}{3} + 1 + 1 = \dfrac{\pi^3}{3} + 2 - 12.335$

(l) $\left[\dfrac{5x^3}{3}\right]_1^3 - 45 - \dfrac{5}{3} = 43.333$

(m) $\left[-\dfrac{x^2}{2}\right]_0^1 = \dfrac{1}{2}$

(n) $\left[-\dfrac{1}{x}\right]_1^3 = -\dfrac{1}{3} - \left(-\dfrac{1}{1}\right) = \dfrac{2}{3}$

(o) $\left[-\dfrac{\cos 3t}{3}\right]_0^1 = \left(-\dfrac{\cos 3}{3}\right) - \left(-\dfrac{\cos 0}{3}\right)$

$= 0.6633.$

Exercises 22.4

1. (a) $\left[\dfrac{7x^2}{2}\right]_{-1}^{1} = 0$

(b) $\left[\dfrac{7x^2}{2}\right]_{0}^{3} = \dfrac{63}{2} = 31.5$

(c) $\left[\dfrac{x^3}{3}\right]_{-1}^{1} = \dfrac{2}{3}$

Exercises 22.5

1. 26.
2. Evaluate $\int_{-1}^{2}(4 - x^2)\,dx$ and $\int_{2}^{4}(4 - x^2)\,dx$ separately to get 9 and $-10\frac{2}{3}$ respectively. Total area is $19\frac{2}{3}$.
3. $[e^{2t}/2]_1^4 = e^8/2 - e^2/2 = 1486.78.$
4. $[\ln t]_1^5 = \ln 5 = 1.609.$
5. $[\frac{2}{3}x^{3/2}]_1^2 = \frac{2}{3}2^{3/2} - \frac{2}{3} = 1.219.$

Solutions to Chapter 23

Exercises 23.1

1. (a) Discrete; (b) continuous; (c) discrete;
   (d) continuous; (e) continuous; (f) discrete;
   (g) continuous; (h) continuous; (i) discrete.
2. 1970   3
   1971   0
   1972   1
   1973   4
   1974   7
   1975   8
   1976   2
   Total   25

Exercises 23.2

1. (a) See Table 1; (b) $\frac{1}{30} = 0.033$;
   (c) $\frac{2}{30} = 0.067$;
   (d) $\frac{4}{30} = 0.133 = 13.3\%$.
2. (a) 5499.4 hours will appear in class
   5000–5499; (b) 5499.8 hours will appear in
   class 5500–5999; (c) the class boundaries are
   4999.5–5499.5, 5499.5–5999.5, 5999.5–
   6499.5, 6499.5–6999.5; (d) all classes have
   the same width, $5499.5 - 4999.5 = 500$
   hours; (e) the midpoints are 5249.5, 5749.5,
   6249.5, 6749.5.
3. See Table 2.
4. The second class boundaries are
   20.585–20.615. The class width is
   $20.615 - 20.585 = 0.03$.

Table 1

Age	Tally	Frequency
15–19	ЖІ	6
20–24	|||	3
25–29	||||	4
30–34	||||	4
35–39	||	2
40–44	||	2
45–49	||||	4
50–54		0
55–59	||	2
60–64			1
65–69		0
70–74			1
75–79			1
		30

Table 2

Mark	Frequency
0–9	0
10–19	4
20–29	3
30–39	2
40–49	6
50–59	15
60–69	10
70–79	5
80–89	11
90–99	4
	60

Exercises 23.3

1.

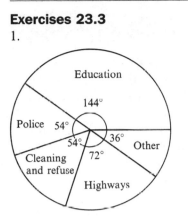

(a) Pie chart showing local council expenditure;

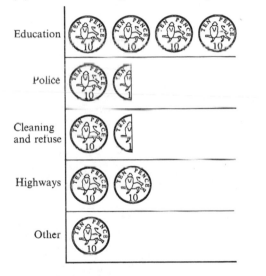

(b) pictogram showing how each £1 spent by a local council is distributed

2.

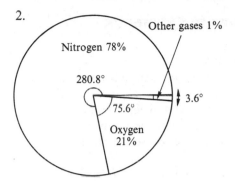

Pie chart showing composition of the atmosphere

3. (a) The class boundaries are 0–99.5, 99.5–199.5, 199.5–299.5, 299.5–399.5, 399.5–499.5; (b) a claim for 199.75 would be placed in the class 200–299;

(c) Distribution of claim amounts

Solutions to Chapter 24

Exercises 24.2

1. Mean $= \bar{x} = \frac{50}{9} = 5.556$, median $= 5$, mode $= 5$.
2. Mean $= \frac{262}{7} = 37.4$. The large, atypical value of 256 distorts the mean, it may be more appropriate to use the median or mode here.
3. Mean $= \frac{498}{7} = 71.1\%$, median $= 68\%$.
4. (a) $x_1 + x_2 + x_3 + x_4$; (b) $x_1 + x_2 + x_3 + x_4 + x_5 + x_6 + x_7$; (c) $(x_1 - 3)^2 + (x_2 - 3)^2 + (x_3 - 3)^2$; (d) $(2 - x_1)^2 + (2 - x_2)^2 + (2 - x_3)^2 + (2 - x_4)^2$.
5. (a) $\sum_{i=3}^{6} x_i$; (b) $\sum_{i=1}^{3} (x_i - 1)$.
6. Mean $= 7.57/6 = 1.26$, median $= 1.32$, no mode.
7. £32080.
8.

Midpoint x_i	Frequency f_i	$f_i \times x_i$
20.57	3	61.71
20.60	6	123.60
20.63	8	165.04
20.66	3	61.98
Total	20	412.33

Mean $= \bar{x} = 412.33/20 = 20.62$ (2 d.p.).

Exercises 24.3

1. (a) $\bar{x} = 20$, variance $= 0$, standard deviation $= 0$; (b) $\bar{x} = 20$, variance $= 12.5$, standard deviation $= 3.54$; (c) $\bar{x} = 20$, variance $= 200$, standard deviation $= 14.14$. The large standard deviation of the final set reflects the widely spread values.
2. Mean $= 5$, variance $= 9.2$, standard deviation $= 3.033$.
3. Jane's mean $= 51.4$, Tony's mean $= 51.6$. Jane's standard deviation $= 12.77$, Tony's standard deviation $= 17.40$. The students have a very close mean mark but Jane has achieved a more consistent set of results.
4. (a) See Table; (b) The modal class is 45–49, this is the most common class; (c) mean $= \frac{2865}{50} = 57.3$, variance $= \frac{10720.50}{50} = 214.41$, standard deviation $= 14.64$.

Class	f	Midpoint x_i	$f_i \times x_i$	$x_i - \bar{x}$	$(x_i - \bar{x})^2$	$f_i(x_i - \bar{x})^2$
20–24	2	22	44	−35.3	1246.09	2492.18
25–29	0	27	0	−30.3	918.09	0.0
30–34	1	32	32	−25.3	640.09	640.09
35–39	1	37	37	−20.3	412.09	412.09
40–44	3	42	126	−15.3	234.09	702.27
45–49	11	47	517	−10.3	106.09	1166.99
50–54	4	52	208	−5.3	28.09	112.36
55–59	3	57	171	−0.3	0.09	0.27
60–64	10	62	620	4.7	22.09	220.9
65–69	5	67	335	9.7	94.09	470.45
70–74	5	72	360	14.7	216.09	1080.45
75–79	1	77	77	19.7	388.09	388.09
80–84	3	82	246	24.7	610.09	1830.27
85–89	0	87	0	29.7	882.09	0.0
90–94	1	92	92	34.7	1204.09	1204.09
Total	50		2865			10720.50

Table for Exercises 24.3, 4(a)

Solutions to Chapter 25

Exercises 25.2
1. (a) $\frac{5}{6}$; (b) 0; (c) 1/3; (d) 0.
2. $\frac{12}{13}$.
3. (a) $\frac{3}{10}$; (b) $\frac{3}{10}$; (c) $\frac{7}{10}$.
4. There are 36 possibilities in all. (a) $\frac{1}{36}$; (b) 0;
   (c) 0; (d) $\frac{1}{36}$; (e) $\frac{26}{36}$ or $\frac{13}{18}$.
5. $\frac{1}{30}$.

Exercises 25.3
1. (a) $\frac{143}{150}$; (b) $\frac{7}{150}$.
2. 0.95.
3. 89760.

Exercises 25.4
1. $\frac{3}{6} \times \frac{1}{2} = \frac{1}{4}$.
2. $\frac{1}{52} \times \frac{1}{52} = \frac{1}{2704} = 0.000370$.
3. $\left(\frac{1}{2}\right)^8 = 0.0039$.
4. $\left(\frac{1}{6}\right)^4 = 0.000772$.
5. (a) 0.5; (b) 0.25.

Index